PADDLE WHEELS
AND
PISTOLS

by

IRVIN ANTHONY

Author of
DOWN TO THE SEA IN SHIPS

ILLUSTRATED *and* DECORATED
by MANNING DeV. LEE
and
LYLE JUSTIS

Both steamers were under a full head of steam

✤ CONTENTS ✤

Contents

✤ ILLUSTRATIONS ✤

ILLUSTRATIONS

PADDLE WHEELS
AND
PISTOLS

CHAPTER ONE

THE INDIANS AND THE FATHER OF WATERS

IAGOO the adventurer rose. For this he had lived through months of weary travel. His eyes glinted proudly as he looked upon the feathers of the hawk and eagle, upon the bearskins and the ring of painted tribesmen. Mondamin, the god who gave his body to the Ojibways, that from it their people might have corn, had brought him his triumph. Godlike, he, too, had given the strength of his body for his people, these of his blood. Fine, large words sprang to his tongue, eager to be spoken at the coming of his hour.

Behind him the wall of white sandstone lifted sheer into the dusk. On either hand it stretched along the river until the twilight lost it. Spring was in the night, the cool spring of northern woods and upper reaches of the Great Water. It whispered in the new leafage of the forest, the song of it rising from the deep-voiced river. The mellow music swept on to the cave mouth that pierced the face of the sandstone cliff, the lilt of running water falling upon the ears of the Ojibway council before the white smoke of a new fire.

Iagoo smelled the strong odor of raw dyes and freshly cured skins. The tang of wood smoke grew about him, mingled with the fumes of tobacco at the passing of the pipe. Beside him, where he stood, an old chief arranged upon the ground a knife, a bowl, a wampum belt. No one else stirred. When he had finished he settled back to listen, for he was the living record of his tribe; all that passed would be stored in his mind, not an accent lost, not a figure of speech changed. Iagoo stooped, took up the knife from the ground and offered it to the old chief, who held it pointed toward the sky. Iagoo touched the keen blade lightly with his finger tips and spoke.

"May paddle fail me, and canoe upturn, may all weapons have edges for my body, and this knife its point at my throat, if I saw not these things of which I speak."

The encircling faces lightened at the solemn words. The old chief hung the knife upon the long pole he had ready. Iagoo dared not talk loosely when so bound to tell the truth. The firelight leaped and slashed at the cave-mouth shadows. The river voice grew. The lean, travel-worn brave took up next the belt of wampum, and holding it his hands began again.

"Brothers, with this belt I remove grief and sorrow from your hearts. I mourn with you for those who have gone with the great whiteness. Into the far northern skyland they have passed, toward joy from grief. I come to you out of the south where the peace of the

14

great whiteness never goes. I bring spring to your hearts. I bathe your heads and bodies that your spirits may be renewed like the sun's. I warm the seats about the council fire, that you may take your ease. This belt is thus a token of my heart: for the rest, my words must serve."

At first he spoke of things they knew. While he had been to the Great Water, they had been living among the pines with bears astir about them and overhead the flight of wild geese. The Great Water had more branches than the loftiest pine, greater strength than any bear, and in its travel out-distanced the most daring goose. Then Iagoo took up his story boldly.

He began at an afternoon before the snows. Low clouds above the river bluff. A biting chill in the forest air. The lightly launched canoe. The song rising under the clouds as he paddled and the rush of the current. The light craft carried southward carelessly as a fallen leaf, or a wild rice stalk. A night sheltered under a clay bank. Another in the deep grass of a river point. A morning of white frost, and bitter chill that reached the bone. A prayer to Great Mystery for an omen, and from dawn to dark, the wedge of flying geese passing over head, pointing the way.

That sign was good and Iagoo fastened a little fetish of bundled twigs into the bow of his canoe. Great Mystery thus knew that Iagoo trusted the omen of the geese flight. There was no time to hunt and very little to sleep. Weary flesh complained. Spirit sagged, but the geese were flying and Iagoo dared not loiter.

Thighs grew used to the thrust of the thwart. Shoulders knotted no longer. The long paddle swung hour after hour. A few measures of ground corn, a short snatch of sleep: they were the portion of the voyager. For the rest, his way lay south.

New sights. A long house of bark and saplings. A strange tent of long poles and sagging hides. A herd of bear-like creatures running upon little feet, their backs hunched as they drank from the river. A great wind that snagged the canoe upon a stranded tree-root. Eleven chains of rapids, running wild and white. Strange lands. Unknown perils. Warmer weather. Tall birds with thin legs, short, thick birds with bills like a swelling gourd. White swans silent as the dead. Owls at night screaming dolorously, like dying men. Huge water monsters with terrible snouts swirled and champed in the mud under the banks. The musk of huge animals hanging heavy upon the air. Great cats come down upon the headlands to fish. Warmer weather still. White branches of the sycamores against the sky. Live-oaks, willows, cottonwoods. Blue and gilded sunfish aflash within the clouded water. Mosquitoes humming. Swamps sick and dark, full of evil smells. Always the Great Water. From its heart round, blister-like masses of water boiled and raised. Unseen strength. Chaos, confusion of its spirit, but always the canoe so far from home going still down into the south, away from the great whiteness of the frozen north over the swan road, down the Great Water.

An arching sky-bowl above a flat country. Wide

16

water edged by swampy points of land. A sluggish current. Ahead the narrowing of banks and at last the branching of the Great Water where fingers of swampy earth bearing cane and stunted trees combed the current. Which to take? Would any bear back toward the White Rock? Did the Great Water end under a rock wall, swallowed by the earth as some said?

The weather clouded, the wind drew ahead. Bravery and wisdom led toward the right bank. The wind cry grew louder. Seas broke, white and menacing, larger than the riffle swell in the up-river rapids. The sun was hazed and nearly down. The night would be wild and there would be rain. Rain! The long line of it drew like a veil across the river stretches. To halt for the night; there was sense. The right bank was gained. The pushing prow drove out upon the river mud. There was shelter behind the point. Lone, tired, the hero stooped at the bow of his craft. An arrow sped past, mis-sent upon its errand. Gone fatigue. Gone fear of the river and the night coming down. The canoe went afloat willingly. The paddle dug deep. Strength sent from Great Mystery over-tried it. It snapped at the loom, at the widening of the blade. Alone with the wind-lashed water, the flying spume, the stinging rain. Alone with the night. There were evil charms upon the happening, a foul plight, whatever it led to.

Down the wind whistled the spirits of the dead. Storm devils rode there clutching at the frail guards of the canoe, storm devils seeking to devour. Bark-thin between its withed frames the craft lived, rearing like

17

the prairie people's ponies, while there was nothing to do but scoop out the gaining water and pray that the devils might not prove altogether evil. Death, alone, under the night was of demons only, demons who raised such evil storms and assailed an adventurer's canoe.

At last the craft slackened, then dawdled. It found its way into the blackness. The rain fell in torrents. The seas battered less and less at each yard the canoe travelled. The storm devils had it in their grip, taking it where they would in the night. The wind song was more shrill, but its power was gone. The dark was peopled, on every hand shadows lurked. The wind crooned to a rustling sound as of myriad whispering. Death lights aswing in the blackness appeared, devil glims rising and falling. The skiff slid toward them. Its prow grated upon a beach. In a dream, bound by a dread spell the adventurer landed. A great people were gathered there, bowed under the rain, and those with flares went to and fro, long shadows tilting with the torchlight so raw upon their paint.

"I, Iagoo, the Ojibway walked up through the bowing people toward the rumble of a great rattle. When I had walked past the people, as many as the rain drops, I came upon the rear rank of the shamans, the medicine men. One lifted his axe to stop me, but I had come out of the night under a spell, and none could do me harm. Stone-axe, nor club, nor keen edge opposed me, and I went on to the second line of chieftains. There I knelt, bowed down by who can tell what power, bowed in the rain under the chanting of the shamans, the hum of the rattle and the blackness of the night.

"And when I lifted up my eyes I knew I had come upon the end of the earth. Before me stood a huge tree, not green, but brown and almost leafless, like one of our own just before the snows. My heart knew it for the tree of the shamans. My blood turned to ice. For there are four great trees in all the world and the greatest of them is that sere and ragged cypress by the mouth of the Great Water at the end of the earth. Above it sit all the spirits of the dead and it is continually awhisper with their thoughts. So much our shamans taught me. They and their teachings were the length of the world away and I crouched in the wind and the rain, listening to unknown words, whether of man or spirit I could not tell.

"The talk went on for a long time. They sent some to see the path by which I came. When they found the canoe they returned to tell that I used no paddle, for they found none in it. All looked on me with wonder, and if I read their manner rightly, with fear. That I saw in their faces. There was no need of words. The rain stopped before they had finished their council. The wind drew toward the trail of the sun. Two shamans raised me, took me into a thick-walled lodge, dried me and dressed me with every honor. When we had eaten I no longer feared, for spirits feast not in the firelight.

"In the morning I was taken to a knoll behind the first tree of the world. Pointing fingers showed the swinging path of the Great River, until afar off I caught

19

a glimpse of shining water, wider than any lake, round like the sky, and very bright with its shining. It was there the Great Water, the father of all streams ended, there in an expanse that reached on down the world to skyland. The air from it was dank like the breath of the salt lick on a dewy morning, and of great life to the nostril."

Iagoo stooped then and took up the bowl. It bore a blue stripe of peace which all knew at a glance. He held it up before their eyes.

"A token! Many more had I. A robe of fine feathers, good squaw work, a knife of rocky wood, and other gifts I brought away, but the hazards of the geese-road took them; they are all gone from me but this. From the *Chitimacha* it has come to our northland, from the end of the Great Water to *Imnijiska,* our white rock, it has journeyed. I, Iagoo, have endured, have bent the back and tried the heart. No small courage carried this token out of the land of the *Chitimacha* to the council fire of my fathers, the braver Ojibways. To the counsel of our shamans, the blessing of Great Mystery give I thanks. Through them I have seen what never man of my tribe has seen, so like Mondamin bring I wisdom to our people. I have spoken."

The embers flickered and died, like those of that other fire by the storied tree of the southland. The voice of the Father of Waters boomed out of the cataract all the night through as it raced past the village of Kaposia, where lived Iagoo who knew its language, and who had looked upon the shining water beyond its end.

✤✤✤✤✤✤✤✤✤✤✤✤✤✤✤✤✤✤✤✤✤✤✤✤✤✤✤✤✤✤✤✤

CHAPTER TWO

FROM THE GOLDEN SOUTH

THE moon set in a cloud bank. The sky was overcast and darkened quickly. A caravel dipped to the gulf swell, slogging her forefoot into the seas with a soughing rush, lifting her bow high out of the crests. She had been clear against the moon path, her square rigged foremast, her long lateen spars on the main and mizzen, but even the sweep of her hull was lost with the moon. The night had turned murky, black, without shadows, and ahead an unknown way. Somewhere on the port bow lay the low coast. The caravel had left Vera Cruz, place of the true cross, three weeks behind along that shore. Then, the course had been northerly, but the land had reached out ahead until, to keep the beach abreast, she steered east. All day her people had seen the purple haze of it, but by night there was no land; only a cloud bank, and a threatened change of weather.

Forward, in the very eyes of her, stood two seamen. One had a trumpet with a pewter mouthpiece and battered brass coils turned green by the sea. The other stood plucking at his black moustache. Both looked ahead with a habitual sea stare, looked and saw nothing, but stood their watch like men. A white crest broke, hissed along the caravel's side, and slid away. There followed the lurch and swing of the bow. The wind was light and on their beam.

Overhead the sails flapped and filled. A low Spanish voice spoke hoarsely from somewhere aft, a speech rheumy with sea damp. Feet pattered, ropes dragged along the deck, thumping a little, only to thud a moment later into new coils. The yards came round, blocks clattered and creaked. The big foresail had been trimmed to catch the puffy, baffling wind. There was soon quiet again. All the watch were idle save for the two lookouts standing silent, fixed above the water that broke and blackened everywhere before them.

At the stern, where the caravel rose, and rose again into two towers, one above another, dim figures stirred. High above the stern flared three lanterns, shining upon rusty mail, upon sea cloaks, high boots and broad rimmed hats worn at a rakish angle. It was there the officers lived and ruled. The lieutenant had seen the moon set in a cloud bank. Was there to be wind? Was the weather about to turn dirty? Were they too close in with the coast? Were his lookouts awake? He was uneasy. He spoke quietly to another standing by, a thick fellow who rolled easily to the vessel's lurch, as if he were of the sea itself from birth. He vanished down the ladder-like stairs, crossed the lower level, and plunged into the low waist of the vessel. There were men, many of them, vague blobs of shadow bundled into corners among the ropes, hidden beside the gun carriages which bore brass breeches stamped with the arms of the King of Spain, long deck guns of small calibre. There he shook a seaman, tumbled him up, and drove him along forward to the heavy ropes that held the foremast in place. The loose shrouds swung to his weight. Clumsily the fellow lay aloft. The wind blew away the scuffling sounds of his climbing. He disappeared. The lieutenant had added another pair of eyes to those already looking ahead and seeing nothing. The caravel had a mast head lookout.

The air was off the land, a dank breath that chilled and bore to the ship an earthy odor, a muddy stench as of swamps and green life. It was an air to worry better seamen than the lieutenant. The threat of the sea, that was the day's work, but this poking about unknown coasts in the dark, it was asking too much of a man. There were no Christians on those shores, no hope for shipwrecked men. The sea seemed honest by comparison. The commander—the lieutenant shrugged his shoulders, sniffed in his peculiar fashion and walked doggedly up and down—commanders were always like that. They only thought of—how could you tell of what commanders thought? They never thought of their lieutenants anyway, and sometimes not even of their ships. Sea room, that was what the caravel needed, not this hugging of land.

At midnight, the men of the watch shuffled below. There were no bells, nor any greetings between the watches. The proud gentlemen in the towering stern cabins would not be disturbed. Only the sailormen had to forego their sleep to work the ship. None but the new officer of the watch stirred aft. For half an hour he let the ship alone. Then, when he had shown courtesy to the relieved lieutenant by changing nothing about her during that time, he took charge in his own right and ordered the topsails taken in. The wind had freshened a little, but not enough to make shortening sail necessary, still, the topsail came in. There were no orders as to carrying sail and the officer felt cautious. It was the land breath. It was anything the crew wanted to think it. To the officer himself it seemed sound seamanship. The night was unhealthy.

Usually, when there is a change of weather due, it begins between twelve and four. Every seaman knows that, and the commander was a good seaman. He left his carven cabin and comfortable berth, and when he reached the deck he felt uneasy, even as his officer of the watch had before him, even as the crew, moving about in the waist, peering over the bulwarks into the night. He ordered a sounding taken with the heavy lead. Like a plummet it went over; the coils followed smoothly until they snapped taut with the weight of the lead. There was no bottom beneath them at thirty fathoms. The lead line hung straight up and down along side. This was new. Where was the ship? Shoals had reached far off from the coast all the voyage before. The crew scooped up a bucket of water from overside. They hauled in and washed the lead. It was sea practice to so treat it though it had touched nothing. The thickset boatswain dipped it into the wooden bucket. He soughed it up and down in the dark. The water spattered about the deck, splashed upon him, striking him in the mouth. He stopped washing and stood bolt upright. He was Portuguese, but he swore a Spanish oath under his breath, then he said:

"Agua fresca, Caramba!"

The seamen at his shoulder heard. Fresh water miles from the land? The officer of the watch was told. The commander learned it in turn. All the people of the caravel knew the night to be unnatural. The air was filled with bogies, the witch breath from the shore was unholy. The commander trod his own side of the deck, and did not go below as he had intended. His officers kept out of his way. Who ever heard of thirty fathoms of fresh water, and no bottom? The responsibility was his. It was he who was taking this caravel into unknown danger. Ahead in the dark lay a presence now, inimical no doubt, at least careless of the safety of the explorers. It would be safer to turn and run to good salt water. The land breeze would aid. The caravel would be safe. The commander had but to give the word. Slowly the watch passed as he stood at the rail of the poop, looking forward over his darkened ship.

Behind were the three great lanterns, forward in the forecastle was a swinging glim. At the commander's feet shone the feeble glow of the shrine lamps, where was hung a Spanish *Christus*, the figure of the Savior, sculptured and in color, the cross in dark raw wood. Anxiously he looked along the deck, waiting for the morning. A thick-set figure stole aft and stopped before the shrine. It was the boatswain, and he snatched his cap from his head, bent his knee and said a prayer. The

25

commander thought him a superstitious dog, but in his heart he wondered why not? Was there not an omnipotent hand in all this? The simple sailor was right. The people were no longer stowed about in the corners of the deck. They stood in little groups, talking quietly. A power of evil was set against the caravel. Otherwise, why should the swell come from ahead and the wind blow across the ship? A force of the devil lay there over the bows, waiting in the dark. He could feel the force of the thing, recognize its power on those simple hearts below in the waist. Fear, terror lurked in such miracles. Thirty fathoms, no bottom, and fresh water! He shook himself out of it, and took a turn about the deck. Three hours would give him full daylight. Three hours of waiting! He would play the man. The east would grey in an hour dead ahead. A fig for retreat. Onward!

At dawn the commander found his caravel in a great stretch of open water. Muddy grey, it ran out toward the horizon, and there lay land. With full light the wind came about until it was fair to drive the ship toward the beach. Soon he raised a lower spit to the right, and with land on either bow the Spaniard sailed in proudly. The water was still fresh, but the bottom had shoaled to seven fathoms and the blessed daylight helped them to gain the river stretches. There they spied a great Indian town, more like those of the south than he had expected. There was no mystery now save the mighty river itself. What a stream it must be that they could sail all night in it almost out of sight of land.

The commander had never heard of such a current. He thrilled to his discovery and his pride grew, forgetting his night fears.

Once off the town he anchored, and as the ship settled back upon her cables the current rippled and sang about the bows in eager wavelets. An Indian came off in a large canoe, an Indian clad in white from head to foot, aloof, suggesting the might of the river. He warned them by gesture to depart and went back to his town, but the commander was no longer timid. A river, even so great as this, was but a river, Indians were Indians, and trade should be trade. The caravel stayed six weeks. Her people dealt with the large town and with forty other hamlets, but they could not breast the current to climb the river, so when their trade was ended they sailed away.

In the south they had separated from Cortez at his finding of Vera Cruz. They had left unwillingly, feeling that all opportunity was left behind, but now they had discovered this giant among water courses. There was news worth taking home. It paid for the night of false omen. The expedition was a success, but who would believe them when they told of their find?

The commander named the great stream the *Rio de Santo Espiritu*. Perhaps he thought of his boatswain when he sought for that name. Perhaps he saw that squat, thick-set fellow bowed beneath the shrine at the break of the caravel's quarter deck, he of the hard word and cruel fist who prayed in the dark. Perhaps—but who can sound the heart of a proud don of a noble

27

nation? He tasted no name so good, that is certain—River of the Holy Spirit.

The commander was Alonzo de Pineda, the first European to sight the Father of Waters. His report was true. He was brave, but his caravel could not breast the stream, so he loitered, and marvelled, and at last went away humbled, bested by the great water. This was in 1519.

Nine years passed before five ships came to try their strength against the river. They too were Spanish and their commander was Pámphilo de Narváez. On their way they rested, first at Santo Domingo and then at the isle of Cuba. When at last they came upon Florida they raised the flag of their king and claimed the land for Spain.

Then de Narváez started off with nearly six hundred men and eighty horses. He had with him Cabeza de Vaca, a clever man upon whom to lean. As a leader de Narváez was more fitted to take a military expedition over the roads of Europe than to undertake the wilderness. He divided his forces, taking with him three hundred men. To each man was given two pounds of biscuit and half a pound of bacon. On this they travelled fifteen days without finding other food than the palmito, a dwarf fan palm, such as the Spaniards of Andalusia were familiar with. They were very loath to lose sight of the sea, and parties were sent out to keep contact with the beach. It was a waste of energy and a cruel business. These outsiders had often to travel through shallow lagoons with the water to the knee and

The party went on under the great trees

oyster shells underfoot, which cut them badly. They lost their first horseman by drowning, in crossing a river that was swift and deep. This they regretted, but they cooked the horse and so provided themselves with a meal beyond their short rations.

The Indians came to them, many of them, playing on flutes of reed. They offered guides, and the party went on under the great trees so much larger than those of Spain. They found walnut, laurel and alligator trees, cedars, evergreen oaks and pines. The soil was sandy, the growth thick. Moss hung from the trees in festoons and ribbons. The sun was lost in foliage. It was June, but the nights were cold, and the days damp. At last they turned their backs upon the sea. Cabeza de Vaca discovered a trail of Indians travelling in force. Occasionally armed savages were spied journeying at the rear. They would not come close. In eight days the cavalcade had plunged well to the west and the Indians attacked. They were repulsed, but the Spaniards dared not travel further until they learned more of their enemies. Fever came upon them. The Indians killed some. Scouting parties were sent out, some to find the sea, some to come up with the rebuffed Indians. De Narváez found the land too much for him, he became sick. His scouts returned with news that depressed. The bays and creeks were too large to ford and ran well inland. Cabeza de Vaca himself reported that progress was very difficult, so difficult that success had moved far off.

The pride of Spain was very low. Southern forests

had no kindness for pomp. No Indians could be captured to do the heavy work. Food ran short, shelter could be gotten only by work, and de Narváez did not know how to lift the morale of his people so they would undertake the necessary labor. The horses, the glory of the expedition, proved useless, and soon turned into very sorry nags with every rib showing. There was not a plume left in the country save those the Indians wore as they glided from tree trunk to tree trunk, keeping watch upon the intruders. Pumpkins and maize were ripening, but the Spaniards dared not try to gather them. When they worked the Indians were sure to kill any cut off from the main body, and thereby crippled, the body of adventurers found themselves in sorry sort.

The Indians were very apt as archers. Their bows were as thick as a man's arm and eleven or twelve palms long. Their arrows were made of straight canes and very hard. Even good armor was of no avail against them. There were those of the Spaniards who had seen two red oaks, each the thickness of a leg, pierced from side to side by arrows. Cabeza de Vaca saw an elm tree penetrated to the depth of a span. The fire was accurate at two hundred paces. Even mail armor, of woven chain, only splintered the arrow shaft, which passed through, giving a frightful wound. By day and night the savages harassed the Europeans. There was no peace, nor security and sickness grew.

At last de Narváez called all to him in single council asking how to leave the miserable country. One third of the men were unable to put foot to the ground. All agreed there remained no chance to escape save to build boats. There seemed no one able to do that, nor was there any iron, nor forge to work it in, nor tow, nor resin, nor rigging. In this extremity rose Don Theodoro, a Greek, worthy to be a descendant of Ulysses, a cunning man who could work. With deerskins and reeds he created a bellows. Spurs, stirrups and the iron of cross bows were forged into tools and nails. On the fourth of September, under the Greek's direction, the

men began to work and on the twentieth they had built five boats. For caulking they used the fibre of the *palmito*. From the tail and manes of the horses were made ropes and rigging. Shirts were sacrificed to be sewn into sails. On every third day a horse was killed and used to feed those who worked on the boats, and the sick. One of the most difficult needs was to find stones in that sandy country, to serve as anchors. Nothing was wasted. Even the skins of the horses were flayed from their legs, entire, and tanned to make water bottles for the voyage. The Indians gave them no peace. In a single day they shot ten men sent to gather shell fish for the camp. In spite of their armor, the arrows transfixed their bodies from side to side. Forty men died from disease before the boats were finished, not counting those the Indians killed.

When the boats were launched on the twenty-second of September, only one horse was left uneaten. De Narváez took command of one boat and the rest were divided among his officers, the one falling to Cabeza de Vaca having told off to it forty-nine men. When she was loaded, not over a span of the gunwales remained above water, and no one in her dared move. There were no seamen aboard her, nor did any one have an idea of navigation.

The Greek, Don Theodoro, realizing the slight chance the boats had, chose rather to trust to the mercies of the Indians than to take the hazards of such a flight. He went readily as a hostage among them, and all that was ever seen of him after was his dagger, which found its way into Spanish hands eleven years later.

The five boats worked along the coast southwesterly.

Thirty days they sailed and rowed with their food nearly gone. Their water bottles had rotted. Occasionally they took fish from the island tribes. At times, too, it was possible to land and make a fire wherewith to parch their maize, the only food they had. The daily allowance was cut to half a handful a day. In the middle of one afternoon the boat of Cabeza de Vaca rounded a point of land and saw far across a broad river a cape opposite. It was the beginning of the end of the command of de Narváez. The north wind blew them out into the gulf. The stream seemed to be in freshet, for it toyed with them for three days. The boats were separated and re-united, tossed carelessly by heavy seas, mocked by signal smokes of the Indians ashore. One morning they were together again. De Narváez ordered the boats to follow him, but he had taken into his boat the healthiest men of the expedition, and the rest could not keep up. The men of Cabeza de Vaca could not hold the pace, which de Vaca reported to the commander and asked for further orders. De Narváez replied that it was no longer a time in which one should command another. Each should look after his own life. Setting the example he departed with his boat, leaving the rest to make what they might of the conditions. They scattered at the will of the north wind and the night, so that by the second day the brave de Vaca was left alone.

He steered until sunset, but still in the grip of the current they made little way. When the sun went down it was cold in the open boats. Not five men of all his command were then able to stand, and nightfall quieted

DeVaca Steered. He Could Not Sleep for Sorrow

all but de Vaca and one other. Two hours after dark this other, who had served as master of the boat, gave over and slipped down saying he thought he was about to die. De Vaca steered and could not sleep for sorrow. Good men lay stretched in collapse at his feet. The boat was a living horror, so close to death that it was ghoulish. Near the dawn de Vaca heard the tumbling of the sea, which, since the coast was low, roared loudly. He wakened the master, and together, with the last of their strength, they put out an oar apiece and pulled to keep the boat off shore until sunrise. They were too feeble, or the boat was too close in, for a wave pitch-poled her on to the beach, flinging her almost clear of the water. Some of the men crawled ashore, others were too far gone and died in the undertow. De Vaca got the survivors into a ravine of what seemed to be an island, where he made a fire, parched some maize and revived his men as best he could. This was on the sixth of November, 1528.

De Narváez and his boat vanished, another of the five washed in upside down, one foundered, and nothing was ever known of the fourth. That of de Vaca was broken in its rough landing through the surf, so that it sank at once when launched again and had to be abandoned. What was worse they lost everything but the clothes they stood in. The mystery of the great river had scorned the stupidity of so poor a leader. De Narváez was not the man to face its vast stretch reaching into an unknown land. Shrinking, hesitation, illness, weakness, none of these had any merit in the face of the Father of Waters. Inscrutable, merciless, the river reached out into the gulf and snuffed out the little argosy as easily as it might have flooded an ant hill, or swamped the leafy fleet that autumn whirled down from the trees to its waters.

With the loss of the boats the men broke into little groups distributed over all the length of the coast from the mouth of the Mississippi to the west. De Vaca failed to hold his crew together. It was as if fate worked upon them. Some were shot at once by the Indians. Some were made into slaves. Four set off to reach Spanish settlements on the southern or Mexican shores. Two of them were drowned in crossing a river and the other two were taken, tortured and possibly eaten by the Indians. Two others set off upon their own, and when one died the other, in dire extremity, was forced to live upon his body until he too found his end among the Indians. The country was harsh enough to its natives and impossible to the stranded men. Every other year, when the walnut trees bore, the Indians lived for a time on the nuts. They saved even the bones of fish they caught, which, ground to a powder, they used to support life as an iron ration to be used only in extremity. In season they ate prickly pears, keeping the rinds, which, when the pears were gone, they powdered and ate. Lizards were eaten, roots were devoured eagerly, only occasionally did they take deer or come up with wild cattle. There was no surplus to feed the unlucky Spaniards. No wonder they were so badly used that few could live.

Dorantes made his way finally into Mexico, married
and had a large family. One of his sons, Balthasar, was
sometime king's treasurer at Vera Cruz. His father
never returned to Spain.

Castillo reached Mexico, where he married and was
given half the rents of the Indian town of Tehuacan.
He, too, never went home to the old world.

Estevancio, an Arabian black, the first negro to reach
the new world, lived among the Indians proudly. They
respected his unusual color. A taking fellow. "Little
Steve" his name meant, perhaps more for his endear-
ing qualities than his size. The savages thought him a
great medicine man and fought each other for his pos-

session. He was more primitive than the Spaniards and more enduring. But, alas, somewhere he got an Indian rattle and this he carried among the Zuñi Indians. The rattle was the badge of an Indian medicine man. Fearing the black, and the magic of the rattle, or doubting how he might have come into its possession, in a moment of indecision, they shot him as he was seeking seven fabled cities of splendid grandeur.

Only one man returned to Spain, the resourceful Cabeza de Vaca. He made his way to the south till he met Christians. Pitying his sufferings and the wrack and ruin of his expedition they dealt tenderly with him until he was fully restored. Two months he rested in Mexico, then embarking in a ship for Spain, he was wrecked. A man who had endured so much could not be held in check. A perfect home fever was devouring him. On Palm Sunday, 1537, he sailed again, this time in a leaky ship, which he quit for another, and so in his third ship he reached Havana. Storm swept him to Bermuda, twenty-nine days saw him reach the Azores. Even then ill luck still followed him, for he fell in with a French corsair, but fate was kind and he escaped into the convoy of a galleon. The commander of the galleon, learning who the ship carried, undertook its escort in memory of the ill fated de Narváez. "Now that you have escaped, follow me, and do not leave me that I may, with God's help, deliver you in Spain," he commanded. So de Vaca at last sailed in under the hills of Lisbon, to the safety of its bowl-like harbor, the only survivor of the six hundred, who sailed to seize and hold Florida. For the rest, except the few in Mexico, the gulf seas and the Father of Waters, sang their dirge under the quiet skies of the new world.

To his king, de Vaca preferred a memorandum of all that had occurred. He had come home from his adventures without a gold piece, but he was the only one to escape so far from Florida. He was received well at court. In his day every tavern rang with rumors of the Spanish conquests in the Americas. He rode high upon this interest of his countrymen, but even when things were at the gayest the memory of that great river would come upon him. His heart was full of thanksgiving and honor. His step was proud. He had struggled in the name of Spanish arms, but his strongest impulse was to go back, to attack, to learn the country to the depth of his desire, to conquer the mystery of the great river that reminded him so constantly of the fulness of his failure. However, he found favor at court and went no more to Florida. Instead he was sent as governor, captain general, and adelantado of the provinces of Rio de la Plata. Once there, his districts got out of hand, rebelled and he was returned to Spain a prisoner. He was banished to Africa for eight years but was finally acquitted.

CHAPTER THREE

AND OUT OF PRIDE

IT could only have come true in Seville. There, in the older quarters, the house of the Moor and the Christian touched. The Giralda was a younger bell tower then, but it had already served the Mussulmans. In the famous palace Alcazar, amid the Moorish colonnades, moved and lived his Christian majesty, Charles the Fifth, and before the new cathedral were erected the columns from a razed mosque. Did not ten choir boys celebrate the festival of Corpus Christi within that church by a solemn dance with castanets, like dervishes brought in from a Mohammedan market place? From the ancient, Torre del Oro, a decagonal tower on the water-side, Spanish sentries looked across the narrow Guadalquivir upon the new village of gypsies. Strange ships from Greece and Africa anchored below its slim height, and mingled with square, bluff bows from France and England. The golden sun fired all the gay colors with its strong light. A city of the south, of strange contradictions and motley ideas. Yes, anything could have come true in Seville even the home coming of Hernando.

Hernando was of good family, but straitly poor. There was no wealth for him in Spain, and without wealth his family could bring him little preferment at

42

the court of Charles the Fifth, but it did give him a chance to go out with Pedrárias Dávila to the Indies of the Ocean Sea. In 1519 he sailed in the second expedition to Darien, and learned some little of the Americas. After that he explored Guatemala and found in Yucatan what the Indians were like, and what could be won from them, and how a little hunger could not kill a man.

His apprenticeship over, Pedrárias sent him out with Pizarro. He had acquired some little fame by then, for three hundred volunteers sailed under the command of Hernando de Soto. He still had nothing but his sword and buckler, his courage and some little experience, but he led his three hundred into the kingdom of Peru. The Indian nation crumpled before their attacks and when their king Atahualpa was captured he was placed in the care of de Soto who found him a gallant figure in defeat and grew friendly with him. Pizarro had succeeded and his reward was as great as his hopes; the wealth of Peru exceeded his wildest dreams. Hernando de Soto was poor no longer. He had not left Seville in vain. Surrounded by stacked gold he had stood in the strong room of the Incas as Victor, about him the spoils of war.

While Hernando was absent from Caxamalca, Pizarro brooded over what should be done with Atahualpa. A dethroned king was dangerous. The Spanish party was but a handful in the midst of enemies. When Hernando returned to Caxamalca he found the captive king had been strangled under orders of Pizarro. De

43

Soto felt he had lost a friend. His honor had been touched, since the native ruler had been officially left to his care. A more reckless man might have murdered Pizarro, but Hernando had wisdom. The memory of the stacked wealth of the Incas' treasure chamber was strong within him. Loftily, as became his pride of blood, he parted with Pizarro, drew his share of plunder, and carried it home to Spain. One hundred eighty thousand cruzados he brought with him to Seville, the city where anything came true.

The king borrowed from him and in the end paid again. Three gentlemen who had come with him from Peru attended him at court, joined by a fourth from Seville. Each brought fifteen thousand cruzados with him. Hernando had learned the value of money in his days of poverty, yet he spent largely that he might show himself to advantage at court. He employed a superintendent of household, an usher, pages, equerry, chamberlain, footmen and all the servants necessary to the establishment of a gentleman. Then, to add to the wonder of his fairy tale, the restored gentleman married Doña Ysabel, the daughter of his old patron Pedrárias Dávila, Count of Puñonrostro. Finally, Charles the Fifth made him Governor of the Island of Cuba, and adelantado of Florida with the title of Marquis which concluded the drama of the improbable that came true in Seville.

Scarcely had his concession been granted than another traveller arrived from the Indies of the Ocean

Sea. He brought no wealth to Spain, but only the broken tale of six hundred men lost by a great river in larger Florida, for the Spanish then had no concept of Florida as merely a peninsula, but gave the name to the lands all along the Gulf of Mexico, and inland to the limits of their ken. Humbly, he pointed to himself, as the sole survivor of the expedition, but his ill fortune abroad followed him to Seville. Hernando de Soto had been made adelantado, and the king was involved with him by a loan. Cabeza de Vaca had no presence he could bring to undo the royal act. He left a written relation of his adventures with the king, and accepting this last blow of fortune as he might, took up his residence in the court city. He had sworn not to divulge certain things he had seen, and this reticence was interpreted by those who met him as a secret of wealth. He hoped thereby to keep himself before the king. When his relatives asked him if they should sell their property and go to Florida with Hernando he assumed a mysterious air and advised his kinsmen that it might indeed be worth while. When Hernando de Soto wanted him to go back with the new expedition he would not go because the adelantado would not reimburse him for a ship he wanted to buy for the expedition. When his relatives remarked upon his unwillingness to go, he told them he hoped to receive another government, which may have been true, for he was shortly made adelantado of the Province of Rio de la Plata. To the end he refused to break his oath

and talk about the wealth of Florida, which secrecy served him well, and yet aided the expedition of Hernando de Soto.

Recruits flocked in, many of them selling all they had to undertake the affair. Nuno de Tobár, Luis Moscoso and Juan Lobillo, the three who had been in Peru with Hernando, leaped at the opportunity. The fourth of the gentlemen who had attended the leader at court, Juan de Añasco, was made comptroller. In those days soldiers of fortune were plentiful and what better symbol of fortune could a man seek than one hundred eighty thousand cruzados from Peru. Portuguese,

46

Genoese, Sardinians as well as Spaniards, both of Castile and Andalusia, as well as of more remote districts, came quickly forward. The final sifting of those who were to go took place at San Lucar at the mouth of the Guadalquivir, fifty miles from Seville. There the adelantado commanded a muster. The Portuguese reported in polished armor, but the Castilians were very showy in silk over silk, pink and slashed. The adelantado considered this a poor beginning for men going into the new world he knew, yet the bearded faces were good to look upon, the men fit and proper. The muster was postponed until the next day with orders that every man should appear with his arms. The Portuguese were in good point, but the Castilians appeared in sad and rusty shirts of mail, and lacking in plate armor, as well as presenting but very poor lances. Those whom Hernando liked he took and sent at once on board the seven ships he had ready. So they set sail on Sunday morning and passed to sea over the bar with great festivity, much sounding of trumpets and the firing of many charges of artillery. Six hundred men were bound for the Indies of the Ocean Sea, and then for Florida.

Hernando de Soto was as yet unknown to many of them. Had the three who had been with him in Peru spoken, they might have said that he was dry of speech and inflexible of purpose. Although he liked to know what others thought, and listened to their words, after he had once spoken he did not like to be opposed. He acted always as he thought best, and all

47

bent to his will. There might be errors in such a man but no weakness, providing only that he be wise.

He had begun in the light of experience. In Peru he had seen the dangers of factional fights, so he placed all the Portuguese of his party in one ship. This they had to themselves under a leader of their own blood whom de Soto could trust. So that they might not grumble at their treatment, he saw to it that they had a fast ship, faster than all the others but his own. Thus he sought their loyalty, believing their disinterestedness might, if won to his side, give him their service if he needed it to handle his own countrymen. Thus, in their time, have reasoned the Popes concerning the papal troops, and the Kings of France in choosing Scottish guards to keep their palaces.

When the fleet touched at the Canaries, and at Cuba, he gave his men what chances for pleasure fell their way. By the time they gathered at Havana he had learned his men. Such changes as he thought fit were made. Doña Ysabel, the bride of Don Carlos, and two other ladies said their last good-byes, the ships sailed grandly in fair weather, and the expedition went forth from the little village of eighty houses and the blessings of its priests, to come upon grim reality.

Hernando had one purpose in going to Florida. He wanted gold. His interest in the land was simple in that. All else mattered nothing. To guide him he had the knowledge of the failures of Narváez, whose men had come out of the wilderness through Mexico, and whose adventures were well known to the Spaniards by their relations there. From that effort Hernando had learned there was no great wealth to be found along the gulf coast. He had learned, too, that the coast was difficult travelling and therefore he decided to plunge inland as soon as possible. De Vaca's hints of wealth were vague, but he had said one thing in confidence to the king, which had in turn been confided to Hernando as the royal adelantado. When the explorers found cotton cloth among the Indians they could hope to find gold, silver and precious stones. Nothing else was so definite, and since it had been given in confidence, de Soto respected that and was on the lookout for cotton cloth.

When the ships had landed at what is now the mouth of Charlotte Harbor, not far from Tampa Bay, de Soto showed his hand to the men. He had a camp built at once. About it he had a space cleared of trees that he might use his horsemen to crush the flank of a body of attacking Indians. Felling trees was hardening work, but good training for what the adventurers might expect, and was necessary to give protection. The Spaniards were none too numerous, every man was precious. De Soto trusted the Indians for nothing. Moreover, the Spanish method of exploring was to make the natives do their work. There were irons and chains a plenty. Indians could be captured, put in manacles and made to carry the equipment of the adelantado through their country until the next tribe was met and pressed into service. Chieftains were usually eager to pass them on quickly enough, as a burden well rid of, and often connived with the Spaniards against the tribes of adjoining land to unload the unpleasant task and to be able to resume their own lives undisturbed. The only difficulty was to keep the Indians from fleeing at the first contact. This took time and patience, but Hernando managed skilfully.

The landing in Florida took place on Friday, May thirtieth, 1539, and parties were sent out from the camp at once, mounted, to speedily find Indian carriers. Permanent camps rot men quickly, and Hernando wanted to harden his men, not let them into idleness.

His outriders had not gone far before they came upon a white man, a survivor of the de Narváez force, Juan Ortiz. He cried out to the cavaliers who were charging upon him and his Indian friends, begging for life, torn by joy at the sight of Europeans, and fear that he might fail to halt the horsemen before he was killed. Willingly he joined the new attempt. In all the years he had not been more then ten leagues from where he was found, but he had learned Indian speech and life. For Hernando he was a valuable man, a link with the earlier effort, a man whose counsel could avoid errors, once the privations of his savage life had been relieved, and he was once again more like a Christian than an animal of the wilderness.

Underway, at last, the explorers travelled up the coast to the port of Espiritu Santo. The land was low, marshy, very wet and heavily treed. The Indians kept out of the way. The Spaniards had to do their own carrying. Hernando sent back the ships, and losing no time, plunged inland.

At last they came upon a village and its chief. De Soto took the chief by the hand, led him back into an ambush, talking through Juan Ortiz as interpreter until the Spaniards were all about. Hernando raised his hand, a trumpet sounded and the cavalry charged upon the chief's village. Thirty or forty fell by the lances. The rest escaped into a pond which the Spaniards surrounded all night. The chill of the water, and the lurking death, by cross-bow fire and the lances of horsemen, who waded their beasts into the water and slew any Indians seeking escape, brought about surrender. By four o'clock in the morning all had yielded except twelve principal men who preferred to die where they were. These were gone in after and dragged out by their hair. All were put in chains, loaded with the goods of the expedition and found themselves "doing labor proper to servants," which in the wilderness, and under the pomp and pride of the Spaniards, was a strenuous business.

On the seventeenth of August, 1539, de Soto crossed the trail of de Narváez. The minds of all were filled with foreboding at the memory of ill fortune that wiped out that party. Many wanted to make for the sea, to find a port, to go back to Cuba. Anywhere seemed better than the fetid heat of the forest, and the swamp breath of the chill night that carried fever to them. So shaken were they that Hernando found him-

51

self facing a crisis of leadership. Shortly, he refused
to go back, even though they lost their way in the forest.
Until he had seen with his own eyes the wonders
promised in the land, things that appeared incredible,
he would not retreat a foot. Then he made good his
word, he ordered them all to saddle, united his force
by calling in his scouts, and began a five days' march
straight into the hills. By so doing he entered a country
where maize or Indian corn was plentiful, found higher
and healthier land, and quieted the ghosts of misfortune
by hard work.

Over the winter, progress was slow. The Indian
bearers died or filed off their irons and escaped. Up,
and then down again, to the southward, they marched,
seeking gold, and rumors of gold. By March of 1540
the men were able to face a march of sixty leagues with
their food upon their backs without despair. Men and
horses had been lost, but the spirit was sturdy. All still
looked for the wonders of the land that had been promised to Hernando by the royal tongue of Charles the
Fifth, by the suggestive whispers of de Vaca.

Hernando found it easier to face his enemies the
Indians as time went on. They were very different
from the weak but more highly cultured natives of
Peru. He claimed he was a child of the Sun, come from
his father's abode, and when they asked why he was
come he answered he was going about the country to
seek the greatest prince. He had reached a land of
large towns only two days' journey from the sea. There,
he had as a guide a feminine chief who had carried

about with her a casket of unbored pearls. In boring them the Indians used heat, so ruining the color, and Hernando longed after these jewels unavailingly, for just when the leader had decided upon demanding them of the lady, known as the Cacica of Cutifachi, she disappeared, taking with her her gems. She might have been a symbol to Hernando, the soul of Florida itself. Everywhere he met rumors of wealth beyond, and everywhere he failed to come up with it.

Time rolled by quickly. Another spring came, March, 1541, and the little band of venturers were still on the trail. Things grew no easier and hope much less bright. That very month they came upon a town of Mauilla where in battle and by fire they killed twenty-five hundred Indians and lost eighteen precious men. It was then that Don Carlos was killed. Long might his bride wait among the lowly houses of Havana, going up to the thick-walled church to pray, going down to see the sea rim across the harbor. Low might the head of Hernando de Soto bow in sorrow over the body, thinking of the sorrow of his niece telling beads in lost Havana. But up, up must come the head of Hernando, for he was at war, and such a war.

Indians seemed to wait everywhere—now starting from amid the low palmito or elders, now stepping from behind tall trunks of elm, or walnut trees. Painted and ochred, they were hard to see. Red, white, black, yellow and vermilion, they appeared to have on stockings and doublet. Some wore feathers, others horns on the head, faces blackened, eyes encircled with ver-

53

milion. They fought on the run, shooting rapidly, and making a great noise.

Bit by bit, attempt after attempt, the Indians came against the venturers. To escape them they made for a river they had heard of, greater than any in the country. They had heard so much of that sort of thing they did not regard it seriously. Then they came upon it out of the woods on to a plain where they camped. It was very wide and in freshet. It had overflowed the level land and rose to such a height that only the tops of tall trees were visible. There seemed to be no shore, only more tree tops across the stream, afar off. De Soto advanced and looked long upon it. Whatever else in Florida might be true to its repute, this river was the mightiest of his world. He ordered piraguas built. There were Indians on all sides, but they found the rovers vigilant and so did not close with them. Maize was gotten from such of the Indian villages as pillaging parties could reach.

The Indian corn when ground, mixed with water and boiled was made into the sagamité, of Indians. Made Indian style, in earthenware bowls, where the water was kept boiling by the addition of red hot stones, it was often very dirty and not very appetizing, but it served to keep life in men and enabled them to come and go all across the great central plain. It also gave strength to Hernando's boat builders who sawed planks out of green lumber, just felled, and built the simple vessels with which they crossed the river.

De Soto called the river the Rio Grande and esti-

THEN THEY CAME UPON THE RIVER

mated it to be half a league wide. A man could not be distinguished across it. There were many snags but also many fish. Evil lived not without good. All the party were gotten over in the early morning so that by the time the sun was two hours high the crossing was complete. Thus the calmest part of the day and that with the least wind was made to serve. Once the need for the piraguas was over they were taken apart and the spikes were kept for making others when they should be needed.

Times were better on the west bank. Travel was easier, food was not so scarce. The towns were palisaded and towered. Loopholes served for defense and the stockades were strongly set and heavily built. Yet the tribes were more friendly, men and horses fattened, and Hernando did not travel very far before he decided it would be well to winter. Gold was as elusive as on the day three long years before when they had landed at Charlotte Harbor.

It was time to take stock. After he had wintered he intended to build two boats and send one to Cuba and one to New Spain to give word how the expedition fared, to recruit new men and beasts and to come back up the river that, reunited, they might go on to the westward. The expedition had lost two hundred fifty men and one hundred fifty horses. What the winter might take could not be estimated. The leader felt the moment was rigorous and demanded staunch handling. To keep the men at top vigilance was both necessary and difficult. A soldier was sometimes set to cry out that he saw Indians in order to ascertain how fast the men would hasten to the call. Those late were disciplined. Captive Indians were made to fortify the camp by a palisade at some distance from the cabins that the enemy might not be able to easily set them on fire.

Unfortunately for Hernando, his high-handed treatment of the Indians had commenced to bear fruit. More than once he had cut the right hand off a thief for punishment, or the right hand and nose off several braves to impress their chief with the obedience due to Spanish arms. Wintering on the Rio Grande, he learned that for miles below him, along the river, was nothing but cane brake, great bogs and thick scrubs. In addition the Indians were uniting into large groups, actuated no longer by wonder at the Spaniards' presence, but by rage at harms done and jealousy at the ease of the intruders' life. Hernando saw no way out. He fell ill and had to go to bed, but he sent to a chieftain a request for his appearance. He got this answer which sounded like the crack of doom to the adventurer so nearly at the end of his rope.

"As to what you say as to being the son of the Sun, if you will cause him to dry up the great river, I will believe you: as to the rest, it is not my custom to visit any one, but rather all, of whom I have ever heard, have come to visit me, to serve and obey me and pay me tribute, either voluntarily or by force. If you desire to see me, come where I am; if for peace, I will await you in my town; but neither for you nor for any man, will I set back one foot."

The messenger who brought this news escaped unscathed. Hernando was low of a fever. He sent out a punitive force against the tribe, who murdered the whole tribe: men, women and children. There was no comfort for Hernando in that, for he saw no way to the sea for his sadly reduced organization, and he knew he had led them into his last fight, for life was going out of him, although every day brought a blessed softness of spring to his blood that made him cry out with the sadness of what he knew.

It was not the fashion for a Spanish gentleman to shrink at the touch of death. De Soto lay through the spring days looking out into the sunlight, thinking of Doña Ysabel, now lost to him. It would be sad news for her. He had failed to find the wonders his king had sent him out to learn. He had not been able even to save Don Carlos and send him home alive. Then the fever rose until the world seethed in great waves to his sight, and all rose and fell before his eyes until he closed them to shut out the reeling vision, and buried his head with his world awhirl.

Conscious, at last, that death was at his elbow, he sent for his officers. They had learned each other well in those three years, both in action and in enforced idleness. Vasconcelos, the commander of the Portuguese, was still with him, faithful and full of courage. Juan de Añasco had never complained at leaving the gay court of Seville for adventure. Louis Moscoso had been as loyal and high-hearted in Florida as in Peru, where the gold madness had first bitten them. Good years of comradeship had passed since then, years of pride and honor and the delight of seeing new land and meeting new obstacles bravely. They came gladly to their chief's cabin at his bidding.

He received them with dignity. He was grey of face and tormented, but the will that would not let him turn back poor as he came was still strong. Pain and weakness had shaken all but that. It only took a little now to end life as became a noble. When he spoke to them his words were grave, but his voice was gentle and well modulated. The King's adelantado would even die well. A thrush raised its full-throated song just beyond the door, the south wind stirred the trees farther off. His comrades listened only to him.

He was about to go into the presence of his God. He was thankful that God saw fit to let him recognize the moment of his death. He was conscious of their love, their loyalty, their long suffering. Florida had taken his life, but he had led as well as in him lay. It was his wish to be relieved of the charge which he held over them, to be forgiven for any ill he had brought upon them. It was the will of God that he should lead them no more. He urged upon them courteously that they elect a governor. It was necessary that they should obey without reservation him whom they elected.

Grave dignity was strong upon him. His officers felt as if he were already detached from them, speaking with an unearthly wisdom. He was a good don. One of them, not one of the most tried three, responded gallantly. There was no speech upon Vasconcelas, or de

Añasco, or Moscoso, their hearts were too full there at the end of his trail. Whom God particularly favored he called first away, were the words. The sorrow of all was deep, but they bowed to the divine will. They stood ready to obey. Whom the adelantado cared to name they would elect as governor as with a single voice, and he should be obeyed to the slightest word.

Hernando de Soto named Luis Moscoso, who was elected by a common voice as governor, and as the adelantado's sickness came strong upon him they left him to what peace he could make with suffering.

The next day, May twenty-first, 1542, Don Hernando de Soto, Governor of Cuba and Adelantado of Florida died. "He was advanced by fortune, in the way she is wont to lead others, that he might fall the greater depth: he died in a land, and at a time, that could afford him little comfort in his illness, when the danger of being no more heard from stared his companions in the face": so reads the chronicle of his end written in Portuguese by the Gentleman of Elvas.

Luis Moscoso was given to leading a gay life, and many thought he would prove an easy commander to hoodwink, but they forgot Hernando had chosen him, and Hernando had wisdom. Finding at council that the general opinion favored standing by the intention of the expedition to still seek gold, the new governor turned from the river and pushed further into the land. He no longer had the services of Juan Ortiz who had died in the winter. The force was small and in poor condition, but he yielded to the general demand, only stopping first to inter the body of his chief.

They tried to keep Hernando's death from the Indians, who had been given to believe that as a son of the Sun he could not die. After three days the body was buried at night. When the Indians inquired for him, having seen him ill, they were told he had ascended to the skies, as he had done on many other occasions, but they were not so easily fooled. They remarked the loose earth at the place of burial. The new governor, fearing that his lie would be discovered, removed the body and weighting the shawls which enwrapped it, with sand, lowered it from a canoe into the Father of Waters by night, tenderly giving the last of the great leader to the keeping of the river. Then, having gotten the signed opinions of all, he entered again upon the search for wealth. Guides misled them, the country steadily grew more miserable, until it would no longer support the party by plundering villages. One hundred fifty leagues they travelled, always westward, and they found no rich cities nor any treasure houses. Their treatment of the Indians grew harsher as their plight became more serious. Some they hanged, others they tortured. All summer they went on.

By October a final council was called and it was decided to give up the land, to return to the Rio Grande, and spend the winter building brigantines. Then they could descend the river and sail along the coast until they reached New Spain, or Mexico as we know it. Many were unhappy at this. The voyage by sea would be hazardous. They had found Indians with cotton cloth which de Vaca had said was a sign of also finding gold, silver and precious stones, but the only jewels they had come upon had been a few turquoise. Of gold and silver they had found none. So, although many wished still to run the peril of their lives, rather than leave Florida poor, none could see any hope of finding wealth, and they returned to the Rio Grande.

Near the mouth of the Arkansas River, upon the shores of the Rio Grande, they captured a stockaded village. There, in December, they undertook the building of their brigantines, at a town called Aminoya. Timber was near at hand, good timber, better for ship building than they had found anywhere else in Florida. All rendered thanks to God for so signal a mercy and accepted it as an omen of their ultimate success. Gone now were the dreams of the cavalcades marching proudly as conquerors into cities of treasure. There were no banners flying, no plumes. The armor that was left them was no longer bright, and their clothes were like those of Indians made of skins or of coarsely woven shawls cut to new service. Deliverance was their prayer, a chance to get out of the wilderness. The Rio Grande looked very beautiful to them, so wide, so deep flowing down to the sea, ready to bear them toward Christians where the new governor might again get his full measure of sleep, where the men might recover from the illness the winter brought, and shake off the awful lethargy that grew out of their weakness from insufficient food.

Hernando de Soto had died, but his wisdom still guided his men. There on the Rio Grande, six months after his death, his foresight made smooth their path. The governor had been well chosen. He went to work boldly with a stout heart. The chains, intended for Indians, the iron in shot, were made into spikes at a furnace which he had set up. De Soto in choosing his men had heeded their abilities, with the result that when he needed them Luis Moscoso had workmen in his band as well as soldiers. There was a Genoese who knew how to set up the frames of the little ships. Four or five Biscayan carpenters undertook to hew out the planks and ribs. A Sardinian and another Genoese who were caulkers, were able to close up the seams with oakum gotten from a plant like hemp which the Spaniards named *enquen*. One cooper they had, and he a man sick to death, thin and unfit for labor, but fifteen days before they were needed he had ready two of the half hogsheads that sailors call quartos, for each of the seven brigantines. Fourteen water butts, built single-handed by a dying man—there was determination. A single Portuguese taught others the sawing, and was responsible for ripping out the straight timbers, the rib-bands, carlines and wales. Hernando had died, but his vision still carried the expedition forward. Never were better men chosen. Never did men live more fully up to the promise of their abilities.

Seven brigantines was a large order. True they were left undecked, planks being furnished at the deck level for the men to walk upon in trimming sail, but even so incomplete, the building of them was a herculean task. Friendly Indians brought in shawls from which the sails were sewn. Rope was made from the bark of the mulberry trees. Anchors were made from the stirrups of the saddles. From January to March the work went on in the cold and wet. Indians attacked from time to time, and in March the river overflowed its banks and stopped all work. The Spaniards took to rafts built of logs, carrying their horses with them. They lived as best they could until the river let them work again. In June the brigantines were finished, but the launching of them seemed hopeless until the river rose providentially and lifted them free so that they were afloat at last.

Twenty-two horses were taken on board. The flesh of the rest was jerked and loaded for food. The governor appointed his captains, who were made to give their word and oath they would obey him until they should reach the land of Christians. He chose the brigantine he liked. The Indians seemed loath to let them go, but Moscoso was fearful of treachery and so refused their every attempt to delay him. The seven brigantines under oars pulled out into the current and bade good-bye to Aminoya, where the men had wrought wonders.

The passage down the river was an endless series of fights. Indians attacked them when they touched the banks, when they anchored carelessly, or even headed them boldly in canoes. The Spaniards had hardly a chance to butcher their horses and cure their meat before the savages were about their ears. The natives realized they were in retreat and as the brigantines neared the sea, attacks grew bolder and the fighting more stubborn.

At last they reached the open sea and were ready to leave the kindly Father of Waters. Having joined in council, there were two opinions. One part of them wished to head boldly for the coast of New Spain, sailing out of sight of land and holding a compass course until they fell on the new coast, thus cutting the distance by a fourth. The other side reminded them they had no compass to steer by, that the brigantines being without decks, and frail, could not hope to outlive a storm in the open sea, and that they would be unable to carry enough water to see them safely through. The more timid were the more numerous and the little band set out to go a-coasting all the way to New Spain.

First, they met heavy water and when they landed safely upon sandy keys the mosquitoes came upon them. They came on the wind. The sails which had been white turned black with them. The men could not pull at the oars for them. Inflamed faces, swollen by the insect torment, looked over the gunwales of the boats, but their owners had suffered much more seriously. Of the mosquitoes they made jests, laughing at each others' appearance, made mirth out of the fire and itch of the stings. Hope was astir in them again, hope that they were really escaping from Florida. Juan de Añasco had once seen a chart, and when they had journeyed some days he directed that they give up following the coast to the west and steer boldly more southward, which they did for a night, and in the morning they saw mountains before them, a low level foreshore, and a river mouth, guarded by a bar where waves broke upon a shoal. Finally they ventured in, found a way past the bar and reached the Rio de Panico. The voyage had taken fifty-two days, almost two months cooped up in small vessels, food low and poor, rest almost impossible, surely a hard passage.

On the tenth day of September the governor led them to the town of Panico. At last they were among Spaniards. Dressed in deer skin clothing, dyed black, consisting of frock, hose and shoes, they went directly to the church to pray, and return thanks for their deliverance. Three hundred and eleven Christians had reached Panico. The townsfolk feasted them and treated them tenderly, mindful of their hardships. There was some little grumbling among the survivors, because at the feasting the people of New Spain sometimes set hidalgo and boor, master and man, to rub elbows at the same time, not looking well to differences of condition.

There was much talk of Florida. All admitted that grapes, walnuts, chestnuts, plums, maize, mulberries might be had for the taking, but beyond that it was a sad land for Christians. In 1544 a report was sent to the king of what had become of Hernando de Soto and the search after riches. It came in time to golden Seville where the brave mourned for a time for the loss of Hernando. It spoke of a Rio Grande, wide, deep, beautiful, to which the survivors owed their lives, but it ignored the fabled wealth. The reticence of de Vaca was understood. There was no need to send out other expeditions; Florida was not Peru, nor did imagining create gold and silver, rubies and emeralds. From the date of that report the Spanish King thought little of Florida. A great country that had done the intrepid Don Hernando to death, but a place without gold, a place where pride, and hard heart, and firm resolve could not lead one to tramp into the glory of golden splendor and the delights of the flesh that attended it. The beautiful river was a failure, it carried down no gold to the sea. Therefore, for Spain, it could build no empire, and in its failure was lost the brave struggle of Hernando de Soto who found the truth about the dream of Florida and died for its espial.

CHAPTER FOUR

FAITH AND CHARITY

IT was Easter Sunday 1646 in the great cathedral of Laon. Folks from far and near thronged the nave and reached about the choir to the wall beyond. Peasants, and somberly dressed burghers, and gentlefolk stood humbly with bowed heads, listening, worshipping. Sunlight came down upon them softened by the glass of tall windows, splendid glass of living colors, mysterious, made by old art in secret. The service reached their hearts. It sweetened the breath of spring that blew still chill over the rocky ridge. The old town straddled the height looking down the slopes sternly, even in the sunlight, but to the people in the cathedral there was only thought of humility, gratitude and a love of peace.

Peace was new to Laon. Since the days of Rome it had turned back invaders. True it had been taken many times, but never rested long in the captor's hands. For centuries, against the barbarians, the Burgundians, the English, it had struggled and only fifty years before it had been attacked by Henry the Fourth. The pride of that folk was just. Their enemies had learned to turn aside from the grim little keep and at last they had peace. They were prosperous. They thanked the good God they were so fortunate and they blessed Him for His care.

Young Jacques Marquette was there with his father, who, like many of Laon, could be scholar or soldier at will. There were five other children in the family, but the father looked upon Jacques with just a shade of worry on his face. The lad was slight, thin of face, fragile as an angel. He stood lost in the great nave, beside his seated father, his eyes lifted to the stone walls of the rose window; second only to that of the great church of Paris. The nuns had taught him that, and so he fixed his eyes upon it wondering, dreaming. He loved Easter Sunday. The priests were radiant in their surplices as they bowed nobly before the altar. The swinging of the censer, the sweet odors that the faint blue smoke carried even down the length of the nave, the tinkling of the bells above the suddenly bowed heads, in which he felt a curious awe, these he loved and his eyes shone mistily. No wonder his father, fearful for him, wondered if another Easter would find him by his side. On the tenth of June he would be ten. He was small, very small for his age.

In spite of his frailty Jacques lived to attend school and if he grew slowly, his mind was eager and his spirit pious. He went to the Jesuits at Nancy and at Pont-á-Mousson to study and so pleased was he with the thought of a religious life that he became a novice in the order of the Jesuits at the age of seventeen. This novitiate was not easy. First came a month of meditation, in which the lad undertook to discipline himself under a director. He thought upon the terrors of hell until he felt in imagination the pangs of the lost. Then, having chosen his path to faith, he was soothed and led to contemplate serenity and heavenly peace. In the month he was given a philosophical background upon which the rest of his work rested. Twenty-four hours a day he wrestled with himself, and his thoughts, until he could see his path, his thoughts ordered at last, his conscience directed.

Two years followed in which the novice perfected his humility and his obedience. He performed the most menial duties, the most repulsive tasks of the sick-room and the hospital. He was sent to beg his bread as a mendicant. At confession he was required to reveal his every instinct, every impulse. He was set to report upon his comrades and they upon him. Thus the novice was measured, and judged, and at the end of the novitiate the Society knew where to send him, what to give him to do, and what success might be expected.

Jacques Marquette showed an ability in languages, an open, friendly mind, a conciliatory disposition and the zealous fire of a good man filled with belief in his work. While not actually a weakling, he was not, like many of his order, a giant for strength, able to endure and triumph. For twelve years he studied, taught, lived the life of the priest and looked forward to his dream, which grew upon him as he worked. When he could he went back into Laon to his own rose window. He had seen other cathedrals, for the Jesuits were free footed; their duties carried them from city to city and country to country. He knew Notre Dame of Paris, and Rheims, where he was stationed for a time. He knew too that they were all expressions of the same zeal that burned in him. Faith in stone they were, faith that stood three, four hundred years, loyal, stalwart, blessed by beauty. What was his life compared to them, his fire to carry the torch of belief to unknown people? Humble of heart, nevertheless the urge grew upon him each year.

The Jesuit priests across the water in New France wrote to Paris telling about the country, its life and that of the Indians. Those religious pioneers expressed themselves simply. There were gallant things in their letters, and sad. Now and then a rarely beautiful description could be found, word sent home of a strange animal, a landscape, or the loving picture of a new mission rising out of the wilderness. These men saw everything, and in writing to their superiors they told everything they saw. These reports Père Marquette devoured eagerly. He marvelled at the strength of body, which he knew he had not, and which his colleagues overseas possessed to such a surprising degree. Sensitive, terribly alive to pleasure and pain, he viewed their hardships in the light of the ruggedness of his spirit, which he knew well, and longed to be beside them in their work. Reading those realities his eyes grew bright and his face wistful as he waited patiently, hoping his superior would read his heart, measure his ability, and use him in New France. Year slipped into year and still he worked. He had no golden dream of ultimate ease. Nothing but desperate work to the end of life waited in the new world. His desire

72

was sincere and his eyes were open. The relations of
the Fathers already at work left him no doubt what
he had to face. His zeal drove him hard, yet he was
a Jesuit. For him there could be only obedience, meanwhile he waited and read hungrily.

The "Relations," as the reports to the Jesuit superiors that interested Père Marquette so greatly were
called, were published in Paris by the renowned
Cramoisy. The annual appearance of the neat little
duodecimo volumes came to be an event. They were
read by the court and so became fashionable. In them
Père Marquette sought out the lives of the martyrs
of his order. Secure in France, working and studying
at Rheims, or Charleville, or Langres, he read of Jean
de Brébeuf, he of strength of will and body as well as
spirit. Through the pages of the Relations of 1649 he
followed the good father to his torture. Captured and
led to the stake, the Indians hung about the priest's
throat a necklace of hot hatchets. Then they seared
his throat and mouth with glowing iron and cut away
his lower lip, to stop his speaking, but Jean de Brébeuf was strong. His speech was silenced but he would
not flinch. Finding they could not break him by his
own pain, they led in another captive, Lalemont, with
strips of bark dipped in pitch tied about his body.
When he saw the plight of his superior, this priest
cried out, "we are made a spectacle to the world, to
angels and to men." They made him fast to a stake
and set fire to the bark in which he was wrapped. The
pain bit him. He flung up his hands in supplication

73

to heaven. Brébeuf was a giant for strength, but Lalemont was supported only by his fervor. His delicate features, his slender form gave no promise of his ability to stand the torture, but after the first horror he did well. Sometimes he was beside himself, but when he could, he offered his suffering as a sacrifice to heaven. The Indians poured boiling water over both of them in savage mockery of the rite of baptism. They cut away strips of flesh from the victims, the torture growing more revolting as the Indians found the prisoners resolute and determined. At last they scalped Brébeuf, and opening his breast rushed upon him in a crowd to drink the blood of so valiant an enemy, thinking thus to add to their own courage. Lalemont was unfastened then, led back to a house and there tortured, till, growing tired of it all they despatched him. Brébeuf had lived only four hours for them but Lalemont endured for nearly seventeen.

Almost every "Relation" contained some such record of suffering, hardships undertaken for the faith, escapes from being crushed in the ice, or frozen, or devoured, or lost, or starved to death. All these Père Marquette read, and as he went about his duties felt the urge to follow in the path that led to such glories. Realities of religion he felt them to be, an opportunity to show man's love of God, his perfect trust, his humility of spirit. They were one with the richness of the sunlight upon the glass of his own cathedral at Laon, with the bold beauty of its six towers, the splendid strength of the carven oxen caught in stone. The

same earnest purpose that had shaped that stone and
moulded that glass had left it to France to carry belief
to the Indians. Père Marquette was of that purpose.
As a boy in the cathedral town, as a man under orders,
he had served God as he thought was his duty, but
his heart surged to go, to break into new fields, to
struggle for souls, in strange languages, amid a savage
people. He had the gift of languages. Why did not
his superiors see him as he knew himself to be? Was
he not wasted in France? Was his heart so much in
need of mellowing? In what was he failing? So he
questioned himself and his superiors until at last it came.

Twelve long years he had worked and in 1666 came
his reward. He was sent to Quebec. He landed there
amid warehouses and dwellings and climbed out of
the lower town by a zig-zag street. A fort, a church,
a hospital, a cemetery, an Ursuline convent and the
house of the Jesuits lay along the cliff. The fortifications boasted a flag pole and a few field pieces, none
too large to command the passage of the river below.
The church was small but brave. If Père Marquette
thought the town crude and rough, he was upon the
threshold of days after which Quebec would seem a
second Paris. The French have been French the world
over. The little outpost, thrown so far into the northland had a struggle to preserve the thousand human
contacts which are all in all to the French, even to
French pioneers. They observed holidays with all the
gayety they could summon, and if their courtesy lacked
the aids they were used to in France, their gifts inter-

changed, the little plays, dances and fêtes were at least helpful in reminding them they were still of good breeding and that somewhere afar off the old world still held its pace.

In less than a month Père Marquette had turned his back upon the commanding little town and had pushed on to actual work at Three Rivers. He was sent there to learn the Montagnais language and perhaps to come to a better understanding of his labors. He had come to Canada with the idea of being a martyr to his faith, but he found the work of his order had passed that first wild sacrifice of good men, and was striking at the real founding of their faith among the red men. The winter shut in, the hard, white winter of the frozen north. He mastered his language as he was later to conquer five other of the Indian dialects. He found them difficult, bristling with syllables upon syllables. Merely mastering the simple forms did not suffice. There were many ways of saying a thing, each using new words which a missionary had to know. Broken speech gave the savages an impression of weakness, of inefficiency. Père Marquette satisfied himself only with complete wisdom, anything less was inadequate for his work.

He found the work hard. To it he brought the detachment of the visionary. In prayer he dedicated himself to the Virgin Mary. In place of the simple zeal to die for the faith he blended an element of romance. With fervor he poured out his soul to the Virgin. He saw her in all loveliness, chivalrously, as a knight might have seen his beloved. His faith was that of the older order of Canadian missionaries: he was fit to play the part of martyr, but denied that, he gave himself to extending the Kingdom of God among a people whose idea of wealth was the possession of several hatchets, a store of corn, a few beads and trinkets and the freedom to rove at large. The life demanded physical strength. He lived it as best he could and the longings of his sensitive heart, which found no place on earth, he sent upwards to the Virgin, finding solace and strength in the skies.

The instructions of the missionaries betrayed intelligence. Gentleness forbade the showing of displeasure at dirty food, indifferently prepared. A priest was told not to undertake paddling unless prepared to paddle all day. This was to prevent the Indians judging him a weakling. When travelling, he was not to keep a canoe waiting for him. He was to lend no Indian any of his clothes unless he was prepared to do without the garment until the end of the journey. It did no harm to admire the children of a village, thereby reaching the hearts of the parents. All of these things were practical, helped to avoid misunderstanding and built up good will between the priests and their Indian helpers. Père Marquette found his gentleness an aid, for it gave him a chance to see rightly. He scarcely needed such rules once he had seen the country at first hand. His physical strength was tried daily by the roughness of existence, but his spiritual insight was sound. He was a good priest.

77

Further and further into the wilderness he was sent. Two years passed before he was sent to the Upper Lakes. He was a missionary by then, tried and true. He had learned to live among the Indians without giving offense by an unwarranted insistence upon cleanliness. Whether he wanted so much, or not, he consumed the portion of sagamité allotted to him. What was more important, he had learned to endure weariness and hardship without giving any visible sign. Among crying babies, barking dogs, young braves hungry for praise, and a whole community eager for attention, but critical of any oversight of decorum, he had learned to do the zealous work of a missionary, not a simple task for anyone.

When he was thirty-one they sent him to the Ottawas at the station of Sault de Ste. Marie. It was a beautiful place. The Sault was a wide rapids. The water leaped down the broad rocky bed and just below, at the foot of the rapids, where the river widened and green islands softened the wild ruggedness of the stream, he built his chapel. The altar cloths were simple, the holy picture not of the best. For decoration he had to use forest boughs. But with them the Indians had Père Marquette, a slight but impressive figure, not tall but boldly upstanding. Gay, friendly, sincerely pious, he baptized eighty children in his humble chapel, besides the dying who often turned to him in their last moments. The life was adventurous and difficult to be so interesting, but there was no promise of anything beyond in such work. Père Marquette mastered six languages that he might truly minister to his flock.

Then he was sent to La Pointe du Saint Esprit. It was while there that he came upon his dream. He was a devout soul, but he was also a Frenchman. He well knew that his order was playing a game to win New France. It was fit and proper, and his pride in the task was both wise and godly and not at war with his humility. As he listened to the talk of his Indians he saw himself in a new light. He knew this lonely spot stood at the edge of the world. Furs came to his post taken by unknown tribes out of new country, about which Frenchmen knew little. Strange stories were current of a people without hair or beard who came from the west to trade with a tribe beyond the Great Lakes. Could they be Chinese? Upon inquiry Père Marquette learned these stories were years old. He learned too of a great water called by the tribesmen Messipi. There was much opportunity to think in the little windswept station upon a lonely harbor, with a low breakwater that moaned of nights when the wind was strong. The great water was a river, so much the travelling Indians could agree upon. With the aid of his six Indian languages Père Marquette got that and wrote home about it. Did it empty into the South Sea, or the Vermilion Sea, as the Spanish called the Gulf of California? Could it turn westward and carry one to China? Did it empty into the Sea of Virginia? All these inquiries he put into his carefully written report.

As if the Indians were aware he was writing home the details of their land, the Sioux tribes drove him out of La Pointe. They gave him fair warning and free passage through their land. The mission fell back to Mackinac Island where there had been a station which had been abandoned. Père Marquette built a third log chapel. The fish were plentiful and the sandy loam grew maize and pumpkins. The winters were terrible but food was easy to get. Père Marquette wrote to his superior Père Dablon of these things, of copper found on the shore, of the chivalry of the Sioux in giving him time to take his mission back to Mackinac before they struck, of a friendship for Louis Joliet.

Louis Joliet was the son of a wagonmaker in the employ of the Company of the Hundred Associates. The Company had large concessions in the fur trade, which it tried to work at great profit. Four thousand colonists were to be brought into Canada by the Associates in return for these concessions from the French Government. Louis Joliet sold them furs, traded for them upon occasion, acted as a guide. He was Canadian born and a rover. He had been to France and came back when he found there was nothing there for him. There was no quicker man in the woods to recognize copper when he met it. His knowledge of Indian tongues was as good as a priest's. Twelve years younger than Père Marquette, he grew to admire the slight, gentle little cleric. He knew him for a man, one who could be talked to, who would listen and understand. They could not meet often, but when they did it was a pleasure to both. Curious, these comradeships of the wilderness, where so much grows upon so slight a basis; a word, a glance, a man's gait, the way he wears his hat, his laughter over a good meal.

There came a time when a new governor came to Canada, Count Frontenac, a man who never knew failure. Dashing, brave, and determined never to give in, to stand all attacks, the Governor heard here and there of the great river and it suited his hopes for New France to learn of it. Joliet was at hand with a reputation as a guide, as a man who came and went at will the whole length of the lakes, where many men went but never came again. "He has both tact and prudence. . . . He has the courage to dread nothing where everything is to be feared."

It was upon Père Marquette, busy with his duties, that Joliet came with orders that were life to both. The priest was to go with Louis Joliet. He was thirty-three, he knew languages, he was friendly. Moreover he had been successful at the Mackinac mission of St. Ignace. His Indians loved him for his kindness. Frontenac in Quebec loved success. Louis Joliet knew only one priest he wished to take with him. Père Dablon, the Superior General of all the Jesuits in New France, knew of Marquette's keen interest in the land to the west, and his desire to carry the faith to the Indians there. There was no other choice in the light of experience and Frontenac heeded experience. With Dablon's approval he had ordered Père Marquette to join Louis Joliet and Joliet had carried the orders himself.

He arrived on the feast of the Virgin, which Marquette looked upon favorably. His daily prayers for her intercession and aid had yielded him this triumph. The two friends felt the honor shown them. They were young men bound upon a magnificent errand. They were going beyond the edge of the world, without shelter, with only the protection that lay in their own strength and they were determined to find the river of the Messipi. Standing so upon the verge of hard-fisted adventure they looked each other in the face and smiled.

On May seventeenth, 1673, they had finished their preparations. Two canoes were ready. Five men waited. Smoked meat and Indian corn were on board. The two leaders had compared notes, Père Marquette relating all he could make out of the reports he got from the Ojibway Indians and, at second hand, of the Illinois, Joliet matching it with local knowledge he had gained on his lonely single-handed journeys, for he had often been near the Messipi. Their combined knowledge they had set down in a map to guide them. Above all, Marquette placed the voyage under the protection of the Holy Virgin, promising that if she granted them the favor of discovering the great river, he would give it the name of the Conception. All this done, the seven men went westward in the two canoes, coasting the northern shores of Lake Michigan.

It was a free life they lived. By day they paddled

steadily. There was no eating between early morning and the evening meal, when they landed, built their fire close to the pulled up canoes, and camped for the night. Soon they reached the lands of the Wild Rice Indians, much to the joy of Père Marquette who welcomed relief from the everlasting diet of corn. The Indians were friendly and very unwilling to see the little party go on toward the great river. Ferocious tribes lived there. A demon lurked in one part of the river whose roar was as great as the thunder's, whose home was in an abyss into which the voyagers would certainly fall. Even should they escape this malicious bogie the waters were full of monsters which would devour both them and their canoe. Legend said, too, that the river flowed into heat that was unendurable. To all this, spoken in friendly council, Joliet but smiled openly. Père Marquette concerned himself more with the spiritual welfare of the Indians at hand than dangers ahead. He stopped long enough to ground them in the essentials of the faith. When they had learned to pray, he took again to his canoe.

Intrepidly they held their course to Green Bay. The Gateway of Death was the name the French gave it, a rough piece of water where the going was unusually dirty and sometimes fatal to careless paddlers. Reaching Fox River they fought their way against the rapids. The canoes were heavy, for they had brought all the rice they dared carry. Wading, and dragging the craft of fragile bark after them, they reached at last the open waters of Lake Winnebago which they

83

crossed and entered the upper river beyond. It was a gentle stream and very beautiful, flowing through open land dotted with heavy growth here and there. Elk and deer looked out upon them as they glided through the endless swale of wild rice. Birds rose in flocks over the marshes on every hand, clouds of them, darkening the sky. Flights of duck and brant wheeled and swung. The wind could not harm the two canoes here. There was no sea; only smooth going and always over the graceful bows waited new sights and sounds. It was a life worth while.

On June seventh they reached the Mascoutins. The Indian town stood high and from it the prairies fell away in all directions. The streams had before flowed to the north and east. They were on the heights of the water shed. Hereafter they could hope to find water which would lead them south and west. The Fox River had ended. There could be no advance until a portage had been made to the upper reaches of the Wisconsin. The early Jesuits had been to the Mascoutins and Père Marquette was pleased to find a cross erected in the center of the rush houses. The Indians were but half Christian; they had decorated the cross with red girdles and cured deer hides, and bows and arrows lay at its foot as a sacrifice to this emblem of the all powerful Manitou of the French. There was no time to quarrel with such mistakes. Père Marquette's heart leaped at faith so triumphant over the gap of years since Père Dablon had come that way.

For three days Indian custom held sway. There was

a great council. Gifts were offered. The tribe was told
the governor of the French had sent out this expedition to find new lands. God had sent Père Marquette
to teach the unknown people of the true faith. Guides
were needed to lead them to the Wisconsin. Therefore help was asked. Joliet was the spokesman. He
spoke as the Indians liked, with a flourish and enthusiasm, using quaint figures of speech and a dramatic
delivery. For three days the explorers rested in the
hands of the Mascoutins. Then they set out and found
at once that the tiny streams did indeed flow toward
the Messipi. For a mile and a half they carried their
canoes through the marshes, then over the prairie and
so to the southward-tending rivers. The way was narrow. Trees crowded upon them, islands were clad with
a maze of wild grape. Marshes opened out and closed
behind. On the points of the river bank, sand bars
shoaled beneath them.

The nights were wonderful. The camp quickly made.
Leaping pennons of flame, the quiet of healthily-tired
muscles, starlight and peace. Savory venison or the
odor of bison meat hung sweet and heavy on the cool
air. Then the blessed evening pipes, and the creeping
on of sleep. At last the quiet of the glowing fire, voyagers at rest, far off wilderness noises, skulking shadows
astir beyond the canoes left upside down upon the
bank. In the morning chill a grey river mist rose all
along the stream before the sun. Veiled life moved
everywhere, wild things welcoming the morning,
vague, dun forms of deer at water, sharp doe prints

beside the blunted, broader tracks of the buck, the dark blot of a bird springing skyward suddenly into the dawn, paling, swallowed in the white fog wisps that hung idly, shutting out the blue. And at the touch of the sun, the voyagers launched their boats gently. Cautious, stepping lightly, they knelt at their task at the paddles. The air was warmer, swiftly the course of the descending current grew clear to the view, glassy water, shading trees, walnut, oak and bass wood and, behind, the bold brow of a wooded bluff. Thoughts of copper, and the fur trade, and the pleasure of the day's adventure for Joliet. Thankfulness to God for having brought him so far, marvel at the many deer spied over the dips and downs. Hopes for new laurels for God and France, wonder at the precious friendship of Joliet, and at last, too, the pulsing thrill at the zest of the unknown, the welcome vigil in the face of danger. These for Père Marquette. He was older. He was more responsible. Confident in his companion, he still felt the trust to rest heavy upon him.

One month from leaving the friendly security of St. Ignace, and the kindly red folk, they saw on their right, broad reaches of flat, grassed land and back of that steep hills. It was the seventeenth of June when they passed suddenly into a perfect waste of waters. They had found the Mississippi. Joliet was happy. He sang. Père Marquette was filled with a joy he could not express. The missionary was a man of excellent spirits, but his speech was always sober if not solemn. He felt too much to have his delight find voice. A huge catfish struck his canoe, startling him. He felt suddenly dwarfed, the new river was too large. It was like all the new world, there was too much of it, the current was too swift, the water too broad. Even the great sky-sweep mocked one.

Memories of the warnings of the friendly Mascoutins rose. Already they had met the great fish that had been promised. At night they landed fearfully to cook their meal. That done, the fire was put out, the seven men retreated to the canoes, paddled some way in the dark, anchored, and leaving one man on guard, slept, cramped as they were. The size of the river bore upon them, striking at their hearts. Every day took them upon a path they might not return. Even Joliet's light spirits were repressed, but Père Marquette remained much the same. Years before he had steeled his soul to seek martyrdom: he would not flinch. His zeal would carry him through. Joliet looked upon the slight little priest and in his turn pondered upon their friendship. There are many paddle strokes to a day, and rest at night was often hard to invite. Many were the hours to give to thought. Two weeks they journeyed so, fearfully, striving to get as far down the river as possible before being checked. Then in the mud of the west bank they discovered foot prints and a path. This was no buffalo trail, no reach of wild cat. They landed and followed it until they saw wigwams and were so close they could hear voices. It was time they broke the charm of the river. They had lived long enough like fugitives bound by its spell, afraid of their very shadows. There was no better time to throw off the fears of the Mascoutins, Indian fears that bound them in spite of themselves. Joliet and Marquette had landed alone for the undertaking. If they did not come back the other five were to go home to St. Ignace. Two friends, they looked each other in the face. Then, as with a single cry, up went their shouts that might bring true the dolorous prophecies of the Mascoutins.

The voices were hushed in the village. All the tribe scrambled for their arms, but Joliet and Marquette stood in full view. Four of their chief men came forward to meet the strangers. Slowly they advanced, measuring these who broke upon them with a shout, holding toward the sun two calumets.

The calumet was a sacred pipe. They were usually carved out of a red stone, finished to a gloss and drilled for a stem. The length of this stem varied as the decoration which went with it. Bright feathers, brilliant gauds, decked its thick stalk. There were calumets for war, savage with scarlet feathers. There were pipes for peace and the council. They were holy to the sun and offered to his godly service by reverent hands. Nothing was quite so sacred to the Indians, and Marquette regarded the advance of the four bearing calumets with much relief.

Still more was he comforted by the sight of their clothes, which were of French cloth. He broke his silence, asked them of what people they were, and received in turn the pipe to smoke and the knowledge

SLOWLY THEY ADVANCED, MEASURING THOSE WHO BROKE UPON THEM

89

they were Illinois. On being led to the chief's tent he
came out to them, straight and proud and naked. Holding his hands before his eyes as if to shield them he
said:

"Frenchmen, how bright the sun shines when you
come to visit us. All our village awaits you, and you
shall enter our wigwams in peace."

Within he had gathered all his tribe and when the
strangers had smoked there again they were taken inland to a larger village where lived the great chief of
all the Illinois. Marquette told him he was a messenger from God, represented the power and glory
of Count Frontenac, and asked for information about
the lower Mississippi. With each turn of his speech
he gave a gift to the chief to emphasize his words. The
chief assured them their presence added flavor to his
tobacco, made the river more calm, the sky deeper
and the earth more rosy. He advised them to give up
the journey down the river, but he made them a
present of a sacred calumet which would carry them
safely among the Indians below, and he gave Père
Marquette his son to go with him as his own.

In the morning the venturers were off again. They
passed painted rocks bearing monsters in red, black
and green, Indian gods. They came upon a torrent of
yellow mud that charged across the calm blue current
of the Mississippi, boiling and surging and bearing
logs, branches and uprooted trees. It nearly swamped
their canoes, but they won through. Soon they began
to see the shores grow marshy and a dense growth

edged them. Its tall stems and light green feathery
foliage glowed in the hazy air, under the languid,
stifling heat of the sun. Mosquitoes came upon them.
They fell in with Indians but the calumet they had
gave them safe passage. Buffalo meat, bear's oil, white
plumes were given them at a feast. Then they went on
until near the mouth of the Arkansas they met Indians
who came upon them from above and below. Bows
were strung, arrows notched to the string, but the
calumet and the coolness of Marquette saved them for
a time. He held his fire even when a few excited Indians flung their war clubs at his boat, for the elders
of the tribe respected the pipe of peace. They were
feasted but their lives were plotted against. The chief
warned them he could not protect them.

The seven among the thousand of their enemies
took council. They could not hope to face their way
past this tribe and live. They were sure the Mississippi
emptied neither into the Atlantic nor the Vermilion
Sea, but into the Gulf of Mexico. If they went on,
with the Indians behind them, they might fall into the
hands of Spaniards and Frontenac would never learn
what they then knew. They turned about and stemmed
the current which had so wonderfully helped them.

The river seemed dark and gloomy now. A heat
haze and thunder clouds hung above it daily. Mid
summer was upon them. Marquette fell ill with dys-
entery. The mosquitoes were even worse than earlier
in the season. The unhealthy exhalations from the
rheumy swamps crowded upon the lungs of all, chill-

ing them at night. All day there was no end to the paddling. Père Marquette invoked the Virgin Mary to help them. Joliet fought with every ounce of his youthful strength to drive the canoes north out of the languid, slothful summer. Weeks ran into weeks, and when they felt their strength was done there was nothing to do but buckle in their belts and with set faces paddle on.

At length they reached the Illinois. They knew their great river now, its strength, its size. They were still puny, still benumbed by their Mississippi, but they had followed it to within ten days of its mouth, by Indian report. Home lay ahead if they could but hold out.

The Illinois was more placid. Its forests had real shade, its banks bison and deer. At an Indian village of seventy-four lodges a chief with a band of young warriors offered to guide them to the "Lake of the Illinois," by which they meant Lake Michigan. Every aid was welcome and the seven readily consented.

Joliet was worn thin, but he was only weary. His blue eyes looked anxiously at that other, the slim priest. He paddled always, but there were hollows in his cheeks and green lights in the eye shadows. A certain leaden, pasty color showed beneath the weather-beaten skin and Joliet feared he might not get his representative of the church home alive. Daily he wondered at the pluck that drove him on.

At the end of September they reached Green Bay. They came upon it with a cry of joy, shrilly feeble under the open dome of heaven, but wrung from hearts that would not give up. Père Marquette remembered the bay's other name and sombrely considered it. Would it prove the Gateway of Death, He paddled as always, but his stroke was very weak, his paddle hardly gripped the water, yet he never pulled so hard. In four months he had travelled more than two thousand, five hundred miles.

At the short portage to the mission of St. François Xavier the seven men used the last of that splendid strength which had brought them all back. Père Marquette could look upon his career as a pioneer with pride. There had been no mistakes in leadership. If he could but stand a little more. They paddled the last bit up to the mission with its cross, its friendliness, the life they had left. As the two battered canoes drew close to the beach and were recognized, Père Marquette, the explorer, departed, but Père Marquette the priest was born again. A little group of Indians came down to the beach hesitantly carrying a child. Their notions of decorum did not fill their need. They approached Père Marquette, not Joliet, and spoke eagerly, broken by grief. The baby was dying, would the Father baptize it? With tears in his eyes, weak, sick almost to faintness, Père Marquette took the baby in his arms at the water's edge and baptized it before death could rob the faith of its soul. He turned to Joliet a face of mingled joy and sorrow. He had arrived in time for this, but pity stabbed his heart. He was still the simple soul who marvelled at the glory of

God at Laon, who felt the pain of the world at his heart there on the mission beach.

All winter they lived there. Père Marquette wrote his journal for his superior. Joliet wrote his for Count Frontenac. Together they made what maps they could of their journey. Père Marquette had been attached permanently to the mission. In the spring the two friends parted. No news of their success had yet reached Quebec. Joliet undertook the journey at the breaking up of winter, carrying his journal and maps, and the son of the chief of the Illinois. With his boatmen he paddled along the shores of Huron, Erie and Ontario. It was rough going in the open lakes. The season was early and the winds high, but he drove on, stopping at Fort Frontenac and so down the St. Lawrence. The river was in freshet with the melting of the snows. The running of the St. Lawrence was a mad careening through white water where the flashing blades of Joliet and his men seemed to pull to no purpose. On they went, pell mell, current borne to the very gates of the city. Then in less than ten miles of the governor the canoe was caught in a small white squall. She capsized. All became the current's plaything. Joliet and one boatman were carried to a rock. The other two and the son of the Illinois were drowned. The report, the maps, the two years of work and gigantic success of their efforts had turned to nothing

There was no need to see Frontenac. The journals to prove his claim to add glory and honor to France lay at the bottom of the river. The governor was not one for excuses. He had enough men of that stamp to deal with. Louis Joliet wrote him a letter. It was the letter of a man.

"I had escaped every peril from the Indians. I had passed forty-two rapids, and was on the point of disembarking, full of joy at the success of so long and difficult an enterprise, when my canoe capsized, after all danger seemed over. I lost two men and my box of papers within sight of the first French settlements, which I had left almost two years before. Nothing remains to me but my life, and the ardent desire to employ it on any service which you may please to direct."

There is no word of sorrow there, no complaint, only a plea for more hard knocks, a clamoring after other risks for the glory of Frontenac. He was sorry for the loss of the Indian boy trusted to his care. The child had been of a sunny disposition, wide awake, a good normal youngster, and his loss hurt Joliet, was an added blow to his pride, but no word of it touched his letter to Frontenac. He was invoking no pity. He was performing an official act with gentlemanly reserve.

"For Fortune it will take its place
Let a man do all he may."

Père Marquette's journal survived, but it was not so full as Joliet's which was written to be official. The good man struggled on at his work. He never recovered fully, but for what of life was left to him he had a new vision of the glory of God. Beside his memory of rich Gothic beauty, of storied glass and stone tracery standing in the silence of old Laon lived a new symbol of God: a great river, mightier than any of the world, along its banks a myriad of voices, and above its waters the miracle of day and night and the openness of the sky dome large enough for heaven itself.

✠✠✠✠✠✠✠✠✠✠✠✠✠✠✠✠✠✠✠✠✠✠✠✠✠✠✠✠✠✠✠✠✠✠

CHAPTER FIVE

LA SALLE—THE KINGLY TOUCH

"IT is as you say, Sieur de la Salle, times are not
golden."

The host of The White Horse spoke softly. He was
not a timid man, but one could never be sure of the
Sieur, who was at times violent.

La Salle lay back in his chair. His great coat was
thrown open, his sword hilt before. He had a hawk
nose, piercing eyes and strong mouth. Grace enough
to play the courtier, and a heavy hand when he found
need, both of these report allowed him.

"Times, Gaspard, are what we will. I am out of Paris
by tonight's coach. Perhaps—for good."

"Sieur!"

There was concern in the word. It was a way La Salle
had; to make a few love him. It was the man. As
Gaspard let his heart speak, a small man swept in confi-
dently. He was dark with a clear skin and a fine face.
He had lost a hand.

"I am looking for Monsieur Cavelier, Sieur de la
Salle."

"At your service, sir," countered La Salle curtly.

He was tired of meeting men. For a month he had
been recruiting, judging merits, planning, fitting all
comers into a pattern within his mind.

98

"The Prince Conty commends him to you and honors me with his favor. He has suggested—"

"He has spoken to me."

There was a long instant while the proud eyes looked his man through and through. It was as if he were probing the fellow's soul, his every turn. There was no faltering, no modesty in his gaze. Man to man they faced one another. Outside, the grooms were at work on the horses for the coaches to Blois and Orleans. The wellshaft creaked, water splashed, sabots clattered on the cobbles in the passage. Within, glasses clinked at the passing of a buttery boy through the room.

"I think we will do well, Henri Tonti, you and I. It is in little things we like each other, is it not?"

Tonti bowed. It pleased the great man to be gracious.

"This New France is a merry dance. Can you live so far from Paris?"

"I am a soldier."

"You are more, Tonti, you are a man, you are my lieutenant. Gaspard, three glasses and my own drink. It would please me if you would join us."

"In New France, Sieur?"

"Faith, no, Gaspard, no. Paris must keep its White Horse in case such as Tonti and I come home again."

The glasses came and a fine white wine was poured into them.

"To China, Tonti, by way of New France, to China with me. To the King then."

"Long live the King, Prince Conty and Sieur de la Salle," answered Tonti.

As they raised their glasses the waiting Gaspard got left behind, so that as they drank they heard a sudden, "Long live the King," followed by a gurgling, choking sound as he guzzled his wine. It was not every day he drank with nobles.

La Salle glanced at Tonti and their eyes exchanged a twinkle of understanding. Upon such small things rest good fellowship; a moment's sudden flash of sympathy, an instant shared. For the noble and his lieutenant Gaspard's choking "Vive Le Roi" began a lifelong friendship.

They were off that very night bumping along the road toward La Rochelle. In the morning they passed through fields of poppies. The color was good to their eyes. It was almost the last of *la belle France* for them for they sailed on the fourteenth of July 1678. La Salle gave Tonti to understand that he had a sanction from Louis XIV to seek the Gulf of Mexico from the north, to explore for France, that the Spaniard might have his pride checked. Wild ideas were about. Famous savage cities were said to be found in the centre of the new world. Great waters were rumored to be there. La Salle dreamed of finding a way toward China Seas, a way for the French to all the wealth of the world, to better peoples than the fabled Indians of the Mexican hills with their sun temples and metal gauds. He talked of these things in the long night watches of the northern Atlantic when they sought warmth in the lee of the deck houses. He made magic of them. To few men did he ever so open himself, and Tonti knew and felt the wonder of it all. Together they blessed their great Sun King. Louis XIV held France in his hand; he sought now to hold the whole of the new world, and who could tell where two like La Salle and Tonti might carry the glory of France. It was two months before they reached Quebec and September had ended before they finished recruiting and were ready to venture out of the colony.

Frontenac, the royal governor, received them as friends. A glance at them told him the stuff of which they were made. The strongly aquiline nose, the cold eyes were offset by a determined bearing and an air of high breeding. Frontenac saw the merit of such agents in a country in which he lacked friends. He would use them. They should have his confidence, he would have their friendship. He showed them a map of America that was very different from that of the present. The Great Lakes bore different names from those of today. Their outlines were even slightly fantastic, although in the main accurate. Lake Ontario was Lake Frontenac, Erie was Lake Conty, Huron was Lake Orleans and Michigan, Lake Dauphin. What is now the city of Kingston, on Lake Ontario, was an outpost named Fort Frontenac. For the rest, to the south and west, there were few names and many patches of unmarked land which the map maker had not even tried to indicate. All the streams that emptied into the Great Lakes were given strange French or Indian names. The great central valley of our nation was an openness left to the cartographer's imagination. To La Salle that map was a lure. He really stood at the edge of the world then. Unknown countries waited just over the horizon; nations not even mapped, land that might be high or low, strange peoples and new ways. Perhaps, somewhere, a lake, a river, would lead on to the east. Advantage, taken boldly, might carry him at last to China.

At home, in Paris, sat the Sun King. He saw his country, France, as a halo about his head, a halo of glory and fame. He looked upon the world as a place for conquest, in which he might carve out new realms, new honors for the sake of his kingship. To be King over a great people was a goodly thing; to be lord over the mightiest, there was a crown worth winning. He had directed a war against Spain and a second against Holland. He dreamed of a third into which he would draw not only Spain and Holland, but almost all of his continent. By these paths he sought honor in Europe, and celebrated his efforts by keeping one of the gayest courts the old world ever beheld.

In America his cause rested in the hands of men like Frontenac. The cold courage of La Salle could bring him realms more vast than his ministers could picture. The fealty of a Tonti could open the unknown to the glory of France. Of this King Louis could not know, but La Salle knew it and Tonti knew it, and Frontenac read it in their eyes as the two looked upon the map of their wanderings yet to be. It was not an idle toast proposed in The White Horse to the health of the King. True, it was the fashion, but it was more, too. It was a symbol of their spirits. For the Sun King they looked upon that map with eager eyes, glad there were no names for the parts to the south and west.

It was with a gentleman's finger that La Salle pointed out what he intended to undertake. White and well kept, the hand, as the finger, gave little promise of the man. He poured out his scheme to Frontenac. As Governor, Frontenac had heard many as promising, but the wilderness had stopped them. This was no walking of hedgerows, nor strolling in vineyards. This northland was a savage thing. There were no coaches, nor any peasantry to drive them. For four months a year trails lay deep in snow, and even when the country was open, men paddled or walked; there was no other way to travel. As a gentleman himself, Frontenac knew that this other could not guess the hardships that lay ahead, yet he felt sure of La Salle in an uncanny way. There was an air of constancy about him, quick and surprising, but steadfast also. It was because of that Frontenac sent him on to Fort Frontenac.

Starting in September, after a few days spent recruiting more men for the journey, La Salle pushed on across the Lake. Frontenac he called it, we have named it Ontario. He had already learned that travel by water was easily the most satisfactory way in the new world. He sailed in a vessel of forty tons toward Niagara. It was his intention to seek a place above the falls where a boat might be built, but the ship he sailed in was lost and he barely saved his men and supplies. He reached Niagara, and leaving his party to winter there, he undertook the journey over the frozen Lake back to

Fort Frontenac. Tonti was left in command while La Salle recruited more men and waited for the end of winter to begin his attempt in earnest. He had already come to feel the size of his undertaking. This land of many lakes would take the best of him. Other men who had tried the riddle of that western country had trusted only to frail canoes or rapid hikes carried forward in headlong fashion. He planned upon his return to use the Lakes as a highway. He would build the first ship those waters had ever seen. It should bear him and his followers into little-known country. Then, when it could carry them no farther toward the unknown, it would serve as a link with Niagara to take back furs from the Indians, to bring out supplies and men. He would name it the *Griffin* and it would be a bond between him and that other brave one, Frontenac, who had the griffin in his coat of arms. So should one gentleman honor another.

The next August saw the new ship built and La Salle back in command. At the launching, they fired three guns, sang the *Te Deum* and carried their hammocks aboard to be out of reach of the savages. In twenty-seven days they had sailed the whole length of Lake Conty (Erie) and Lake Orleans (Huron), in no danger from the Iroquois. It was a master stroke to reach so far into unknown territory without risk to his force. There was no loitering in the service of La Salle. Lazy men soon deserted. When he could, he recaptured them, and returned them to their tasks. He had no apparent fear of their treachery nor did he stoop to humor them.

Like the Sun King, he regarded only the ends he was seeking. He had begun well. So far the maps were not useless. The French had known these lakes for nearly eighty years. Franciscan fathers, of whom there were several in his party, had journeyed over them often. They too, knew the way down into the south and west toward a legend-river which ran away boldly to the south. Striking back into the hills La Salle reached the heart of a country given over to the Illinois Indians. There he built Fort Crévecoeur which he so named because of the troubles and suffering he had met in reaching so far. In this fort he wintered until February 29, 1680.

They were anxious days for the commander. The Illinois Indians were friendly. The Iroquois, who might come against them at any time, were jealous of the French, knowing them as usurpers and invaders at Fort Frontenac and along Lake Frontenac (Ontario). The life was rough. Many of the French were broken by the unusual work and took the first chance that came to desert. Their mutiny was sly. They tried to poison both La Salle and Tonti. Things were at a low state.

Yet the life of the explorers was not entirely unhappy. La Salle, at the court at Versailles, was polite and aloof, dressed in velvet and lace bowing before the mirrors of the Galerie des Glaces in the palace. La Salle, on the Illinois, was painted and plumed to undertake the calumet, or the dance of the pipe of peace. He lived by Indian ways where policy made it necessary, yet he was iron-nerved even in his levity. The moment never swallowed him. The Indians felt his power even as they enjoyed his society. Tonti had become transformed even more than his master. He could not lose the martial swing of his shoulders, nor the square set of his head, but he replaced his lost hand by a hook of steel which was the envy of every similarly maimed venturer ever after and the delight of the Illinois Indians. This, among them, was good medicine and Tonti gained a prestige which was more than flattering.

While they waited for news of the *Griffin,* La Salle grew impatient. He felt he had done nothing. Other white men had seen what he was beholding. The Illinois River was not very close to China. Fortunately he could not know how far apart they really were. From the Indians he heard of the Mississippi. It was the Indian's own name for the mighty legend-river. The Father of Waters was an inviting term. Who could tell where it led or what was in store for the man who could reach it? Delay was fateful. The first two months of the year glided by in winter quarters. The Indians grew suspicious that the French intended to betray them to their enemies the Iroquois. The *Griffin* should have been back with more men and supplies before the ice shut in, yet there was no word of her. La Salle sent off two men to gather news. He saw no more of either men or ship, so he sent off two others he hoped would prove more faithful. The Frenchmen were again on the verge of mutiny. Desperate measures were necessary. The time to act came at the end of February 1680.

La Salle had with him, beside Tonti, a Franciscan

father who was every inch a man, a brave heart named Louis Hennepin. That everything in the way of exploration might not remain at a standstill, he sent this man down the Illinois with two Indians to find the Father of Waters, and explore it to the northward. The southern exploration he reserved for himself. On the last day of February Louis Hennepin set off. Tonti's power with the Indians was as great or greater than his leader's, so La Salle turned over the command of Fort Crévecoeur to his able lieutenant. News must be had of the *Griffin*. La Salle set off on March 2, to seek her.

It was a journey for the bravest only, and La Salle chose four Frenchmen and a faithful Mohegan as fit for the trail. There was no shelter to be looked for in all that reckless journey save what the traveller could build for himself. At the tail end of winter even sturdy log cabins were inadequate to keep out the weather. The party foreswore shelter. The Mohegan might be able to provide them with occasional meat but that was mere chance of the hunter. For food, too, they had to trust to fortune with the snow deep and the land so long ice-locked that even nuts had been swept from the trees by the brave north wind. Worse than either of these, which were after all the lot of all venturers, La Salle encountered snow in places over a man's head. While frozen and crusted, he was able to travel, but a thaw, the first suggestion of a return of spring, turned the snow-depths into quagmires of slush where no progress could be made. At times, in ravines, where the snow had drifted, the six men left the ground and took to the trees like apes, like madmen, determined upon some crazy end. Swinging from tree to tree, dropping breast-deep into wet snow, that drenched them to their skins, emerging upon an ice-sheathed ridge to meet the full blast of the northern wind and bitter cold, they fought their way over their thousand miles.

Two days out they came upon the two men La Salle had sent to learn of the *Griffin*. She had been lost, utterly lost, with her crew and cargo amounting to ten thousand crowns. The expedition was no longer in a dangerous position, it was desperately off. La Salle made his decision quickly, standing there in the wilderness, with his handful of men, already lean and hungry, already strained, and suffering from frost-bite. Proud, aloof, with as cold a spirit as the north wind, he went on. He must reach Montreal. There was no other way the thing could be done. Go on he must, so he shrugged and quietly set off and his five men went with him. What is more he got them through. He did not lose a man, not even crossing the frozen Lake Frontenac (Ontario). This was a leader of spirit alone. None reached his heart but all rested on his wisdom and cunning. He won against storm, winter, Indians, the hazards of travel, the ultimate failure of the expedition even to the threshold of Montreal.

Pinched and battered, at last they made it. At Frontenac La Salle learned he had lost, in the lower St. Lawrence, another ship, in which he owned cargo to the amount of twenty thousand livres. His trade canoes had been wrecked and broken in going from Frontenac to Montreal. Men whom he had sent out to trade along the lake shore to east and west of Fort Frontenac had run away with both canoes and trade goods. Nothing could be learned of any of them. His grant covering this fur trade should have netted him an annual profit of twenty-five thousand livres. Everyone in Canada seemed to have turned against the intrepid adventurer. He had nothing left but his self possession and his ability to see things with clear sight; with these he rose to meet his trouble.

On the western edge of his world waited a small body of his countrymen. They were not very loyal, they were scarcely even ordinarily faithful, but they were his. He had to think for them. Behind them, in the mists of uncertainty waited a great river. In its exploration lay his road to fame. He must return to it. The wedge of the French to be driven into the new continent from the north was a blunted, useless thing. The forces to drive it were spent, their energies dissipated. On top of hardship had been piled financial ruin. Another might have given up. La Salle turned in his tracks. He carried nothing with him but his spirit. At Montreal all things were lost to him. He headed west with all possible speed buoyed up only by the rugged friendship of Frontenac and the largeness of his own soul.

Scarcely was he well on his way than he met messengers from Tonti. There had been a mutiny, the Frenchmen destroying their own fort and carrying off

every thing of value. They were making for the settle··
ments. La Salle, who was then on the shores of Lake
Frontenac (Ontario) set out to head them off. Their
trail crossed his as they fled with their plunder. Some
he killed in the cold wrath of despair, some he sent on
to Frontenac in chains, a most uncomfortable way to
travel in country as pitiless as primitive Canada. For
himself, there was nothing to do, but go on to Fort
Crévecoeur and see the worst. He found the country
devastated. The Iroquois had gone through. Tigers
could not have done more harm. They had slain every·
one they met, red or white. In the whole Illinois coun··
try there was nothing spared. Ash heaps, skeletons,
smashed canoes, rotted caches of Indian corn were all
that were left of the people who had trusted him. His
word was with them; he had plighted his aid. There
was nothing to do but gather in the tribes on left and
right. They must help. La Salle pushed down the Illinois almost to the Mississippi hunting Tonti, perhaps
blaming himself for the harm that had overtaken his
friend. He organized the Indians he met into a unit,
with which to face the Iroquois, that this "scourge of
God upon the wilderness" might be checked on the
west, as they were already by the French in the St.
Lawrence valley. If the French were to explore and
reach the Mississippi they could only do it through the
aid of friendly Indians. If the Iroquois could be made
to face a solid front of tribes they would have less time
to observe the French. Stronger than either of these reasons was the urge of the debt that he owed to the ruined

tribes of the Illinois. La Salle was cold, but he was a gentleman. He organized a confederation. The remaining seventeen villages of the Illinois had retired beyond the Mississippi. Sieur de la Salle intrigued to draw together, with this remainder of the Illinois nation, the chiefs of the Outagami people, the Miamis, whom he induced to forsake the Iroquois, and the Shawnees. This union he brought about by his power over the Indians, which fortunately he never lost. They understood each other, those warring braves with their thousand intrigues and vain speeches, and this lone, proud man whose eye never lowered, nor whose voice failed his bidding.

For the time he had suffered enough. Leaky canoes, shoes made of destroyed cloaks, acorns and withered fruits for food, broken only in fifteen days by the lucky slaying of a deer, had lost their charm for even so brave a man. Travelling by the north he connected with Tonti, who had still with him a few Frenchmen including several Franciscans. They all retreated to Fort Frontenac where plans were at once begun for a new drive upon the Father of Waters.

By December 1681 Sieur de la Salle had chosen a party of fifty-four, including both Indians and white men. His friend Tonti went ahead to prepare for the new drive. He found the Illinois frozen so there was nothing to do but set about building sleds. Upon these the main party would be able to drag canoes, baggage and provisions. Hardly had he finished when the force of La Salle came up. They travelled down the Illinois until they came upon their first sight of the Mississippi. They promptly named it the River Colbert in honor of the Sun King's minister. Floating ice checked their venturing farther for a week which they spent marvelling at this new giant among streams. At last the river cleared and they were off into the rile of liquid mud rolling down to them from ten or twelve days away, if the Indians were to be believed. They dropped down to the Missouri, where they wondered that although this great stream emptied into the River Colbert, it seemed to have grown no larger for the addition of that river. Ten to twelve leagues a day was good progress. When they came to the Ohio they stopped for a rest from the cramped and painful life afloat. Ahead lay forty-two leagues where they could find no camps. The banks were low and marshy and full of thick foam, rushes and walnut trees.

The forty-two leagues past, they came upon their first sight of southern Indians. These were a finely set up people, honest, liberal and gay. La Salle, however, omitted no precaution. When he landed an entrenched redoubt was hastily thrown up, a palisade of fallen trees to prevent surprise. When that was done, the French waited to permit the Indians to recover their confidence. To the first of them that came near, the pipe of peace and presents were offered. Sometimes they were afraid to smoke with the white men. Then the calumet was carried to them by a Mohegan, or an Abnaki until their trust was complete and all were on good terms. Always, behind the progress was the sagacity of La Salle, watching, able at a dangerous moment to meet the Indian in his own style and to win him over to the venturer's side.

The French grew rather jolly on so pleasant a journey. The work was hard, but they were succeeding. Already the frigid winter of the Great Lakes was lost. Jests passed from canoe to canoe. Parrots cackled at the laughter. Chansons were sung lustily, the words rolling out from bearded lips. Snatches of café songs long forgotten on the pavements of Paris came to life. The little fleet was merry as the river carried them to the south, the first men of their kind to travel so far down its course.

The tribes of the river lived in houses of earth and straw with roofs of cane. Within were wooden beds and painted walls. Their temples were reared to the sun and within were interred the bones of their great chieftains. There were fans of white feathers, fine cloth of a dazzling whiteness, plates of copper and an unknown metal, both highly polished. This Indian life quite astonished the French, but no more than the country itself. Palm trees, two kinds of laurel, plums, peach, mulberry, apple and pear trees covered the whole country. Even so early in the season the vines were in blossom. Color was everywhere; compared with the greys and browns of the northland, a very sunburst of them. Food ran low. At times they passed through the land of a hostile tribe. Even when they killed a deer a *michybichy* might rive and tear the meat, carrying off great portions. A michybichy had paws like a wild cat, but larger, skin and tail like a lion, body long and large like a deer's, only more slender, and a head like that of a lynx, though much larger. He was the animal ruler of the region, but dared not attack man. Snakes were to be met, alligators avoided, yet the spirits of the company were good, as the country unfolded more and more wonderfully to their eyes.

April ninth was the day. Down through the delta the little band paddled. The water had turned brackish. In two leagues it turned perfectly salt. The tide was in. Ahead lay the open sea. They landed at once and with all decorum began a ceremony. The Father of Waters had been conquered. They chanted a hymn of the church. A procès verbal, or declaration of their attainment was drawn up and signed by all of them and a volley of musketry was fired in celebration. The arms of France were raised together with a cross. A leaden plate with the names of the discoverers cut upon it was buried beneath the monument. Then, with the men under arms, drawn up in order, Sieur de la Salle read in a loud voice:

"In the name of the most high, mighty, invincible and victorious Prince, Louis the Great, by Grace of God, King of France and Navarre, Fourteenth of that name, this ninth day of April, one thousand six hundred eighty two, in virtue of the commission of his Majesty . . . do now take . . . possession of this country of Louisiana."

When he ended Tonti cried, "Vive le Roi."

"Long live the King," answered the little army.

"Vive Sieur de la Salle," led Tonti again.

When that response had been given every man of the group knew he was face to face with a bad business, but every man felt too that he had done well and would pay for his privilege as the way home might demand. The current checked them every foot of the way. There was no food only some dried meat which they got from the Indians at the mouth of the river. Soon after, perceiving it to be human flesh, the French left it to their Indians. They lived upon alligators and potatoes, wrested corn from an enemy tribe by a two-day fight, overcame fatigue by tightening their belts and going back to their paddles, and league by league won back the distance they had covered so easily when southward bound. La Salle drove them without let-up almost to desperation, but he kept them in hand. He had learned there was no new way to China by the great river, but he saw the glory of France that might rise in the new land. Not a man had been lost on the entire journey. When he reached the Illinois country again he set his men to build Fort St. Louis which he left in command of Tonti, while with all speed he hastened to Quebec to sail for France.

He left behind him honest Tonti, and the rugged friendship of Frontenac. From deerskins and pistols he stepped into the silken garb of a courtier, with a dress rapier at his side. He went to remind the Sun King he had acquired a new realm. Louis now owned the Father of Waters, all rivers that entered it, and all the country watered by them. La Salle pointed out to him that he had in his hands the power to thrust at Spain. The Sun King liked brave men, but La Salle had commanded too long in his own right to be sufficiently humble. He was full of dignity, and his pride was very sound, but in the end he played diplomat well enough.

He asked only that the King hire one hundred men for a year. He would enlist five hundred savages if the King gave him six hundred muskets. For the rest he asked most of all pistols, weapons that could be used well against Spaniards or Indians, heavy enough to shatter breast-plates, of sufficient range to check an Indian archer from shooting quickly, close in. In his letters to the King he had thoughts of eventually overcoming the English and Dutch too, but his first effort was to be against Spain.

Louis XIV liked well to harm Spain. He fitted out four vessels with the supplies La Salle requested, but the men sent were a poor lot indeed with which to carve out an empire. One ship dropped behind and was captured by the Spaniards. Another was wrecked at what was thought to be the mouth of the Mississippi. Beaujeau, commander of the fleet, put the new expedition ashore, sailed away and abandoned La Salle. After trying to find the coastal site for a colony, the brave Frenchman set off to strike his river further inland, connect with Tonti, and attempt colonization on a big scale from the people of Quebec and Montreal.

He had set aside his dream of China and the east until the glory of his King was secure. When that was

HE DROVE HIS MEN MERCILESSLY

117

certain he would yet go seeking a passage to the east. Many things had failed with him at the first attempt. Somehow he would yet turn this early ill luck to account. Then, when French voices rose and fell along the Father of Waters he would be free. Once colonized, the valley would yield profit to La Salle, and glory to France. The Sun King would need him no longer, and he would be free to try again for China.

Item by item he considered his needs, but considering them did not cure the plight of the little body of men. Illness slowed their progress. The country was too big for men on foot, and he had no canoes. Powder and shot ran low. He drove his men mercilessly. Some deserted and despairing of ever getting out of the country joined friendly Indians that they might live. Food, from the very beginning of the journey, had been short. La Salle knew the hopelessness of his fight to the last detail. The only remedy he saw was to go on rapidly.

Somewhere to the north, beyond the torturous canebrakes, where the expedition had to painfully chop its path, waited Fort St. Louis and Tonti. Tonti was a fixed star in La Salle's heavens in those days. Tonti of the White Horse, Tonti who ventured among the savage Iroquois as though they had been friends, Tonti who had made the great voyage to the river mouth and back. La Salle felt sure of Tonti.

In desperate haste he drove his men. The old iciness of his temper was stronger than ever. Where another leader might have broken into tears or oaths, the cavalier only grew silent, his face more set, his eyes harder, his understanding deeper. The brother of Lanctot, one of his men, could not keep up. Delay was fatal. La Salle ordered him to return to their latest camp, and as he obeyed he was set upon and killed by the Indians. Lanctot swore vengeance for his brother, but he kept up with the little body of struggling men. Moranget, La Salle's nephew, was sent to hunt. Lanctot fell upon him with an axe and killed him. When Moranget did not come up with them at the end of the day, La Salle returned to find him. Two eagles circled overhead, watching. Intrepid to the last, he walked into an ambush. Duhault and Lanctot lay in wait and shot him as he passed. This occurred on the nineteenth of March 1687. So ended his hope of reaching Tonti, and his dreams of a passage to China.

There were those in the party, men of iron stuff and the same pitiless, heroic mould as their leader, who resented the murder. One of these, an English buccaneer, took two pistols, and accompanied by a Frenchman went off to the cabin the murderers had built themselves, where they had stored what was left to the expedition of goods and necessities. He asked them for some ammunition and some shirts by way of excuse. Lanctot refused, and the Englishman drew his pistol and killed him. Duhault ran for the cabin, but was stopped by a shot in his side. La Salle's Indians came up and demanded the murderer be finished, which another pistol shot did with all despatch. Tonti sought for his master, undertaking a passage all the way to the Gulf to find him, and that at his own expense, but to no avail. Treachery had finished the matter differently.

La Salle was dead in an unknown grave, beside the Father of Waters but he had won the valley for the Sun King living so daintily in far off Paris, and he had earned the lifelong friendship of as good a man as his Tonti, who never abandoned the great river but lived out his life at the work for *la belle* France.

✢✢✢✢✢✢✢✢✢✢✢✢✢✢✢✢✢✢✢✢✢✢✢✢✢✢✢✢✢✢✢

CHAPTER SIX

ON FORTUNE'S WHEEL

ALL across the Northland the French thrust boldly. Trading posts sprang up quickly, settlements of log cabins buried often in heavy snows, frost-bitten and cold-punished, from which bales of fur were sent out by sledge, or when the ice was gone, by canoe. The northern winters checked the blood, staggering the boldest spirits, but the work went on. The land was free and the French were equal to its trials and disappointments. Iberville attacked the British attempt to monopolize the Hudson Bay trade. Furs were plentiful. The French ousted the English and Canada became a living tribute to the energy of the men of France who had but a boy for a monarch, the great grandson of the Sun King. Quebec was his key to all this new-found wonder.

Yet after the day of Sieur de la Salle the French did not let the Mississippi slip back into mystery. The intrepid explorer had pointed the way out of the fastness where life for three months was necessarily a thing of heroism, only supported at a terrific cost of energy and effort. Lone, wandering men travelled to the south, taking pelts, escaping from the spell of the frozen northern wastes. The river was friendly to them. It was a highway, a protection against attack by Indians. The power of its current and the menace of its shoals were not nearly so malevolent as the breath of the arctic from which, in Canada, there was no escape. Brave French hearts were abroad, seeking out the great river arms. They followed the water where it reached into the mountains of the north. They crossed deserts, afloat, to the lost heights of the far west. Often single-handed they wrung profit from the denizens of the land. By water they came and went. Indians, hard-ships, the terror of the unknown ways all were over-come by the beneficence of the Mississippi. Knowledge grew. Reports came in and furs to substantiate them. Iberville went home out of the north to France to build a new project.

The nation needed a settlement in the south. Always the river led southward. The focus of the whole valley was at the mouth of the river. Iberville, in France, gathered two hundred souls, men, women and children. He was intent on making the river mouth French in spite of all of Europe. The little party of pioneers were put on board ship. Under the convoy of the *François* of fifty guns they were carried across the sea, past the Spaniards of the Indies and so to the river mouth. Iberville came with them. His experience was good, his wisdom sound. Success was for him. The two hun-dred arrived safely. A site for the new settlement was chosen close to the river mouth on the east bank. Virgin land welcomed them, waiting only for hard work to fashion it to their desire. In two short years he founded the first French settlement in the southland, boldly seizing the land he needed. His younger brother remained to build well and truly, when, at the end of that time, Iberville went home to tell the King the mouth of the Mississippi was his. Enterprise and thrift had won it for him. New Orleans was the key to the south, even as Quebec held the north.

The Spaniards in the West Indies and Mexico learned of the bold seizure of their territory. A force at Biloxi on the Gulf protested formally against the presence of the French in their domain, but the two hundred were not easily frightened. Iberville had chosen the site of the town with an eye upon possible defense. Low ramparts were built. The inhabitants were trained to fight. The Spaniards had possessed their opportunity for one hundred and fifty years. The French had realized a dream where they had failed, and put it to service. They were willing to stand by their guns if necessary. In the north, stockade forts held the portages by which voyagers could enter the valley from the Great Lakes. Posts were established along the banks of the river. At the mouth lay the infant city and the valley had become indisputably French by the right of possession.

In 1713 the city had grown to four hundred, but the life was hard and the colonists were low spirited. Four hundred against the wilderness were all too few, and the wilderness was always there, unrelenting, eternally returning to front the brave handful. The Mississippi turned and twisted its way into new channels and utterly disregarded the colonists, even as it had de Nárvaez, Cabeza de Vaca and de Soto. The government of the colony had discarried. All were at a point of disgust which promised ruin for the French outpost when John Law appeared in Paris.

He was a conjurer. He took the one word Mississippi and organized a company in its name. The stock of this company was bought by good and bad Frenchmen alike. Upon the boulevardes of Paris it was the word that carried the thing along. Norman farmers from the land of the poppies flung their savings into the company. Storekeepers and lawyers within the city squinted one eye shrewdly, looked upon the project and trusted their meagre savings to the new stock. The whole scheme, it was claimed, was designed to people Louisiana. This, of course, was nonsense. It was a device to ease the finances of the crown. Never before had the river been used as a loadstone for gold. More wealth than even de Soto, with his memories of Peru, and Inca store houses holding heaped gold, had ever dreamed of had poured into the venture of the new world. The whole French nation shrugged its shoulders and began a game with fate for fortune. All rested upon the single word which was become a formula. Law cried "Mississippi" and France from King to peasant gambled upon the sound.

Up, up went the stock in Paris. Many who plunged knew nothing of the company but its name. Some bought because their neighbors, whose wisdom they trusted, had invested. Man matched man. Courtier matched courtier, and the whole city enjoyed the thrill of the wild gamble. Blood leaped at the goad of hope.

There was one great benefit of the fever. It was necessary to even the temporary success of John Law that the river lands should be peopled. Some were lured by the promises the land was said to hold. Others were kidnapped even as sailors are sometimes shanghied. By one dodge and another seven thousand free men were transported to the colony of New Orleans and with them went seven thousand slaves. There was no return for any. Their presence in the new world was necessary to raise the price of stock and upon the advance of that stock rested the welfare of France. Buy and sell, buy and sell, that was the way of the plan. Perhaps Law saw that there would come a time when prices could go no higher. It was a foolish idea, but while it lasted the mill ground merrily and the poor colonists were the grist. They found themselves utterly deceived. The new world held not a tenth of the promises made them, but they could not return. Idlers, workers, rogues and respectable folk, they found themselves in a drab, river city back of beyond. Heartbroken, the victims met the matter fairly, buckled down to work and New Orleans became a substantial town. Man will live let life do what it may.

In 1720 the top price was reached and in France the bubble burst, and the people who had plunged wildly sat back to look upon their madness with pale faces. John Law fled the country. Many were ruined. More than a dozen drowned themselves in the Seine. Havoc was upon the nation. The magic had gone out of the

THE WHOLE FRENCH NATION BEGAN A GAME WITH FATE FOR FORTUNE

word Mississippi, but from a little outpost of four hundred discouraged, hopeless colonists New Orleans had become a thriving little city where doubtless there were many broken hearts, but where too there was much bravery, much energy, much admiration for the reality of the river, the mere name of which had flung so many into the new world.

Amid suffering and struggle, in a world where life was primitive and work heavy, the French won clear. The administration did what it could. The colony was mostly composed of men. Marriageable young women were brought over under the care of Ursuline nuns and married when they left the ship for the levee. The government of Louis XV saw that they were well launched upon their new life. Each girl was given a trousseau packed in a trunk, wherewith to begin life as a bride in New Orleans. It was thus these young women came to be called, *"filles à la cassette."* Many a happy and prosperous marriage was so made. The approval of the government and the blessing of the church sheltered them. It was all part of a gallant effort to carry into New France a high heart and a determination to make the Mississippi as a dream come true, after the fiasco of finance.

The century hurried on. The nations of Europe fell afighting. France lost Canada, Newfoundland, Arcadia to England. That it might seem friendly to the Dons, an act of good faith, Louisiana and New Orleans were ceded to the Spanish. At least the territory thus escaped the British, which seemed to comfort French pride. The governor of New Orleans had to tell his people that their efforts had gone for nothing. Their mother country had cut them loose. France, imperilled, had been compelled to give up her dreams of foreign conquest. With consternation the colonists found themselves turned over to the government of Madrid. The Spaniards were considerate. They allowed the creoles five years to get used to the idea of being Spanish before a governor was sent out to take over the control. At last, in 1768, he arrived. Even then Madrid chose carefully Antonio d'Ulloa, a man of science and letters, who as a mere boy had been savant enough to be sent with some Frenchmen to measure the meridian at the equator. He was just, brave and not unfriendly, which was a blessing to New Orleans. Some who came after him were sad governors indeed, but the little town bore up as best it could.

It was almost time for the entrance of the incomparable Napoleon. France had never given up being sorry they had ceded Louisiana. The War of Revolution had been fought and the United States emerged from thirteen rebellious colonies into an infant nation. Bribes were offered the Spanish King, negotiations were undertaken between ministers. The name Mississippi was not often heard, but its fate lay in the hands of the adroit politicians eager for power and fame to whom the river meant nothing but a piece to be played upon the chess board of diplomacy. As the sun of Spain paled, France rose to the occasion. Napoleon was winning. An Italian kingdom was found for the only daughter of the Spanish King, made up out of northern provinces conquered by the Bonaparte. On October 1, 1800, Spain gave back Louisiana to the Republic and the Spanish King's daughter became Queen of Tuscany. The Mississippi was again French.

It was a dot in the plans of Napoleon. He was winning. Fourteen armies were molding destiny. His dream stopped nowhere. From Egypt to India, from New Orleans to Quebec, he saw his marshals carving the way. Then a very similar event to that which started the World War took place, only instead of an Archduke, no one less than Emperor Paul of Russia was assassinated. The Russians suspected Napoleon of being concerned in it and the enmity of Russia never after was lost. Massena's invasion of India failed, and the Danish fleet upon which Napoleon had counted as an ally was soundly thrashed by Nelson at Copenhagen. No wonder the great Frenchman wished he could be everywhere and could not feel sure of success unless he commanded in person.

Napoleon had time for everything. Education, finance and law: these he ordered so well in his spare moments that France has left them ever since pretty much as he planned them. He earned the respect of his people as a warrior, yet when he was gone his fame grew and grew. Even today there are no loud voices to be heard when the multitude bare their heads at his tomb. They looked down upon his sarcophagus of rich red stone lying there beside the Seine, an endless stream whose hollow footsteps echo forever, slackening only to marvel reverently at the wonder of the man before they pass on under the golden dome. Napoleon had time for Paris. Napoleon had time for the Mississippi.

He had taken it from Spain. He felt that he had lost India and Egypt. The great British Empire was building before his eyes and France had no navy fit to check them. What was worse, to Napoleon, he had no navy to hold the Mississippi. The British had taken Canada, the Isle Royal, Newfoundland, Arcadia and India. Would they be able to take Louisiana too? Twenty English war vessels were gathered in the Gulf of Mexico. Not even he would be able to carry on such a war without resources greater than he had.

On Easter Sunday 1803 Napoleon announced what he intended. Representatives of the United States had approached him about the matter. They wanted to buy the city of New Orleans. He offered them not only New Orleans but all of Louisiana. Monroe and Livingston, the Americans, had no instructions to treat for so much but they seized boldly at the chance. Fifteen million dollars was the price. Napoleon thus kept Louisiana out of the hands of the English. He raised money for his coming wars which he could plainly foresee were about to begin again. He gained a friend in the United States. Thus he profited by that which he could not have kept by any device. Monroe and Livingston felt that what they had signed "would cause no tears to flow. It will prepare centuries of happiness for innumerable generations of the human race." And Napoleon the Great, what, in addition to making something out of a sure loss, what thought he of it? He reminded the envoys of his great kindness to the United States. As he put his name to the treaties he declared that the holding of Louisiana "assures forever the power of the United States, and I have given England a rival who, sooner or later, will humble her pride." That was the feeling of the proudest man the modern world has ever seen.

On May 22, 1803 the Mississippi was bartered away cheaply for the coinage of a young republic, but its proud waters heeded not even the prophecy of Napoleon. Over them were to dawn wild and eager days. No longer were the winding stretches to be a toy for far away kings. Americans were coming into the valley, mountain men, living like Indians despite their English and Scotch blood, coming down out of the hills. The river was American and the new republic's men were ready for their day. Young blood coursed freely, young blood loved the river with the pledge of its current that every mile should offer something strange, even down to the little city at the mouth. The Mississippi had found a people worthy of it at last, who were willing to become at one with its might, to make themselves a part of its age-old story.

In New Orleans life moved slowly. The official giving over of the river mouth and its great valley was one thing, the actual change in government and administration another. The thirteen states which found themselves so suddenly owners of the land were almost as far from it as Napoleon upon the battlefields of

Europe. A wild and little known mountain range lay between the states and the valley of their great river. Such an obstacle has changed the whole life of a nation; to the United States it was but a temporary bar. The young nation was flushed with the dream of new greatness and smiled upon the unfolding of destiny.

The French flag had only flown for twenty days after the hauling down of the Spanish. For twenty days there had been a feeling of joy and content. The city had taken its time to turn the city over to the French who so promptly parted with it again. It was an old world spot in the midst of the new wilderness. True there were full bayous close at hand. The buzzards and the pelicans throve along the river. Cypress, palmetto and live oaks, moss hung and ancient, shadowed the waters, but in the midst of it the city sat proudly, the key to all the river, the gateway of the South.

The defenses were formidable. Soldiers had built the high, sloping ramparts, good soldiers who knew their trade, notably the Spaniard, Galvez, afterward viceroy of Mexico. On the north, east and west the fortifications girdled the city, and on the south flowed the river taking a scimitar sweep before the wharves of the town. It was a mile across. The city was of narrow lanes, ill drained and pestilential. The streets were named for princes and nobles of the court of France. Truly national, loving gaiety and social life, the people struggled for the polite amenities of life even as their countrymen in Quebec had dared the northern winters to give plays in warehouses and dances where they could.

These French of New Orleans looked beyond the wall toward the river mouth to the Faubourg Sainte Marie. There, three deep, lay the boats of the rivermen, strangers who came and went upon the great stream. They were the only Americans the French knew, and they frightened the good citizens sadly. To think of having such men as rulers shocked them. The men of Faubourg Sainte Marie were rivermen. When ashore they played and they played roughly. The cries of their festivities came up to the city by night, wild, untrammelled, filled with the joy of savage men from the "broad horns" as their boats were called. Uncouth and ribald, the life there savored in no reflection of Paris Boulevards of which dwellers within the ramparts dreamed. The houses of the quarter were mere shanties without any permanence. A tropical quality had stolen into the dwellings of the French. Red tile roofs, steep yet massive, walls pierced pleasingly for balconies and verandas, massive wooden doors, delicate iron work wrought into gates and lattices. The cathedral rose proudly above it all matched only by the grey stone and wide balcony of the Cabildo. All this clashed harshly with the cries of the boatmen beyond the pale, the only Americans the people of New Orleans knew at first hand.

To the anxious city came the civil agent of Napoleon, and upon his heels the news he could scarcely believe himself. No one was more surprised to hear of Napoleon's cession of the city to the United States than M. Laussat, the civil agent of France, who was preparing for a formal ceremony of taking over the city so soon as General Victor should arrive with the French troops to relieve the Spanish garrison. It was to be a proud moment in his life, a glad one for the French population so long left under Spanish authority. The new turn of affairs struck him like a thunder clap. His consternation was sincere. The astonishment and sorrow of the people reflected his own. They had lived under Spanish governors and thought themselves badly off, but this new authority they linked with the maudlin, rough-mannered men of the Faubourg Sainte Marie, and they shuddered at the prospect. It was bad enough to find themselves cast off by their motherland, for a second time to be made pawns in the game of politics, but to fall into the hands of people they could not understand, to become the prey of men like those from the "broad horns" and the sea-going ships that lay three deep along the wharves, that could not be met by a shrug, a roll of the eyes, a flinging of the hands, a long look to heaven. They, the last of all New France, were come to their end. Despair was at their hearts. They saw no way out but to submit. It was all too sudden.

On December 20, 1803 the dreaded change took place. There was a ceremony, but a very different one from which M. Laussat had planned and hoped for. Instead of the impetuous General Victor from France, two Americans came. They were General Wilkinson, commander-in-chief of the army and Claiborne, the new governor, and they were the collective symbol of the new order. Claiborne was already governor of the

136

adjoining Mississippi Territory. He was a southerner, gallant, a former member of Congress, young enough to feel sympathy for these French so rudely abandoned by the great Napoleon. He knew the Latin desire for a spectacle. His duty was clear, but the manner of it was at his command, and he chose to consider the emotions of the brave townsfolk who had carried on so resolutely there in the wilderness, where even the river at times threatened to wipe them out. They lived in houses without cellars. Even their dead could not be given graves below ground, for the water rose everywhere. He respected them, and he mourned with them. His spectacle might prove pathetic, but it should not be heartless. He would do what he had to do gently, as became a man of his ability.

There was only one place in the city to stage the transfer of government, that was the Place d'Armes in the centre of the town at the Cabíldo. It was a place associated in the minds of the populace with civil authority. Besides, the quiet dignity of the massive Cabildo, relieved by its staircases and carved railings, were equal to the impression Claiborne wished to make. He knew he could never understand this foreign people, nor they him, but in all kindliness he undertook the matter gallantly.

A body of American troops was marched in through the open gates. They carried their own stars and stripes, but the music to which they stepped was alternately French and American; strange that early colonial music must have sounded against the full melodies of old

France. Once within the city they were met by the Spanish garrison acting as an escort of honor to lead them to the Cabildo. There Claiborne and Wilkinson presented their credentials to Laussat. Both sides addressed the assembled people and Claiborne was given the keys of the city.

Then, as their governor, he told the people they no longer owed Napoleon any allegiance. He welcomed them to the protection of the United States and pledged them freedom under the new flag. His intent was good, but New Orleans was of no mind to accept promises. The people sought better evidence of faith than that. Then came the moment of pathos.

For twenty short days the tricolor of France had flown over the French city. Twenty days, after forty years of what many felt to be wrongful possession by the Spanish. The twenty days were gone and men of the same blood as the boatmen of the Faubourg Sainte Marie were about to replace the tricolor by their own new flag, which meant nothing to New Orleans. The tricolor was started. As it came down the stars and stripes rose slowly. Midway they met and kissed, flags of the same colors but very different, then the one was mast-headed, but the tricolor was seized by a French officer. Grimly he wrapped it about his body, without bravado or swagger, he marched off sadly to the barracks. And the populace; they filed after him in solemn procession following their beloved tricolor. There was no failure in official politeness. Laussat turned and led the two Americans within the Cabildo. There he ten-

138

dered them a banquet, a lengthy affair of French courses, food for a king served fit for a gourmet. Laussat entertained them in as princely fashion as became his master's pride. Claiborne knew he lacked tact, knew he had not the French nor the ability to join in the city's ways. Course followed course at the table. Proud speeches were made. Ringing toasts were put and taken with filled glasses. Claiborne, however, saw only the bowed heads of the procession following their flag across the public square toward the barracks. He saw them going on and on. All of the dinner they tramped past him out into the jungle of cane and brush, out over the tang of the salt marshes to the river. Beside it they had chosen to settle even as his countrymen to the north had followed its length to choose a home. It was all there was in common between this foreign city and the government that he had brought it, and even that seemed slight enough. He would be honest with them and patient, like the broad, muddy face of the waters just beyond. There was no mistaking their power, there should be no mistaking his. He would win them if he could, but he could not forget the procession of the tricolor.

CHAPTER SEVEN

THE RIVER EMPEROR

A ARON BURR was born to the east, to the masters
of the United States. He had the power that came
from his breeding. Friends flocked to him. Sympathetic,
kindly, with an ability to centre friends about his force-
ful personality, he moved fast to the very front of his
people. They found him a charming fellow, a man who
had fought with honor in the War of Revolution, a
man who thought clearly and who brooked no obstacle.
Any man who has friends, has enemies, and Burr was
no exception. In American politics leaders were either
won to him or their opposition hardened to hatred. He
could be cold as ice, haughty, bitter in denunciation.
Yet he forged ahead, fired by ambition.

In France Napoleon was carrying the palm and the
eagle, inspiring loyalty and winning battles. Burr
dreamed upon his might, glimpsed his power and saw
visions. What cannot a man, gently reared, hope for in
a new country? Napoleon had been one of the consuls,
before he became the first. Why could not he, Burr,
hope to reach individual power? He reached the Vice-
Presidency of the United States easily enough. Jefferson and Hamilton stood in his way, fighting him at
every turn, but that did not greatly disturb him. The
land was young. For the time he kept his dream to

himself. Then at the next election he ran for the Presidency of the United States.

His opponent was Jefferson, the Federalist, and the election was bitterly contested. Hamilton opposed Burr's election in every way. The only thing he could attack was his selfishness and his party. This he did whole-heartedly. The election was close. Burr held his friends magnetically. They believed in him heart and soul; he inspired them with that belief as a great man should. The election was so close it was thrown into the house for final selection and Jefferson was elected President by one vote.

Alexander Hamilton, who had so much to do with Jefferson's success continued to fight Burr in defeat. Burr, wounded to the heart, felt the famous federalist had insulted him, had wrongly used his power and was making a mock of him. He could not be president of the United States, but he could still be a gentleman, according to the European code. He challenged Hamilton to a duel and Hamilton dared not refuse the challenge. The young nation would have read cowardice into that. He was a statesman facing a soldier. Face to face the men met at Weehawken near the stern palisades of the Hudson. Pistols were the chosen weapons, beautiful pistols, long barreled, neatly stocked, usually kept in a box-case for the sole purpose of settling such differences. They were genteel, like the men who used them. They smacked of personal respect and that quick, vital thing, individual honor. Cruel, but precise, was their verdict when appealed to. The ball they shot was nearly a half inch in diameter, the range a few paces. Not infrequently both duellists were killed, and so the verdict of the duel was left unpronounced and the malice perished with the men. Hamilton hated Burr, Burr reciprocated heartily. The handkerchief dropped to signify the moment to fire. Under the stern lift of the river shore, Hamilton fired into the air, but Burr would not be so mocked. He killed Hamilton where he stood. For him the duel was a true satisfaction. His wrongs were righted by a pistol ball, but his day in the east was over. Napoleon might rule France, but Burr would never govern the thirteen states. Hamilton had sacrificed himself for his party, a Federalist to the end. Burr had avenged himself and cleared what he felt to be his besmirched honor. July 11, 1804 was an important day for Aaron Burr. From that moment he became a law unto himself guided by a pillar of fire he alone saw, a hope veiled from the world, but clear to his own heart.

As a politician he knew there was in the west a strong dissatisfaction with the government. A cry of separation reached his ears from time to time, a cry that the east was holding back the development of the great midland for its own benefit, for the enrichment of the seaboard states, for the increase of its own power. In the west, some favored complete independence. Some were willing to return to the British government, others still looked favorably upon the loose adminstration of the Spanish. Clark and Robertson, true Americans, leaned toward this last solution. Daniel Boone, finding the

Mississippi basin too heavily peopled to please him, pushed westward and became himself a Spanish official far up the Missouri. New Madrid, on the Mississippi, peopled by Americans, was quite willing to live under the Spanish flag. Seven of the thirteen states had once claimed as their western boundary the Mississippi, but that had been given up to the federal government. Two hundred million acres thus rested in the hands of officials of the hated Federalist party. Aaron Burr knew all this. His popularity was great in the south and the south was always closer in sympathy with the west and northwest than with the rest of the original colonies. Something might be made of all this unrest. Napoleon was conquering Europe upon a less hopeful foundation. Who knew what might be done with this great flood of immigrants pouring into the open land beyond the states? The hope of success in the east was ended. Burr had given his chance of sitting in the President's chair a death blow by his duel. He would go west, but first he would examine his resources carefully.

Taking stock, the soldier of fortune found he was of expensive tastes with little money to meet them. He considered the possibility of a war with Spain. The seizure of the city of Washington, the kidnapping of the president, a mutiny of the navy in his interest, an intrigue for the support of the British fleet which was to attack New Orleans when he led an army down the river. All these things he thought of as means to raise fortunes which were low. Today these things seem madness, but in 1805 they were all possible. The British minister at Washington was rather pleased with Burr's suggested plan, but he failed to involve his government in England although the matter was seriously considered. With that rebuff Burr gave up all attempts to deal with England. He promptly turned promoter of a colony. Neither land nor people were to be seen. They existed as yet only in the mind of their promoter, but he was persuasive and he had ideas and he talked with a unique charm. When he had convinced himself of the possibilities of the scheme he turned his back upon the east and crossed the mountains.

Wherever he went in the Mississippi valley he found friends. As a democratic leader he drew to him the party that was most powerful beyond the mountains. On the frontier his duel was no stain upon him. Had not Hamilton been a hated Federalist, one of those who handicapped the west? No man so distinguished as Burr had ever come over the mountains and the gratitude of the country knew no bounds. His tongue spoke ill of the east and was music to everyone. His journey was slow, and his thoughts had time to crystallize. Clearly he saw that his sole chance of success lay in daring. Temporarily he would promote a colony, a real flesh and blood proposition of land and settlement. In the neighborhood of Natchez on the Mississippi the Spanish still ruled and evidenced no desire to move. If Burr could have ready a body of well-armed colonists when the probable war between Spain and the United States broke, he would be able to plunge in at once, capture the land and hold it. If the war did not

WHEREVER BURR WENT IN THE MISSISSIPPI VALLEY HE FOUND FRIENDS

come, then the colonists would have to be talked into buying the land waiting for a fit moment to secede from the United States. It was a definite idea, fairly workable, and either way it worked Burr saw himself as head of an empire.

Cincinnati received him well. In Tennessee he found Andrew Jackson, major-general of the militia. In him centered the anti-Spanish sentiment of the district which burned fiercely, a dangerous surge of passion that threatened to rip the valley from end to end in sudden war. Burr was encouraged. James Wilkinson, commander-in-chief of the army, listened to the scheme eagerly. He was then in Spanish pay as well as his country's, but he was ready to throw off both if Burr's scheme promised more reward. Encouraged, the adventurer spoke more fully of his inner dreams than was wise, then he pushed on down the river for New Orleans. There the people liked neither Spain nor America; that should prove a bountiful field. Had the means been Burr's to launch his attempt at an empire boldly, he would have hurled it in the teeth of the world over night. There was no cowardice in him. His dream was modest enough at the time to come true, and not so startlingly treasonable as to shock any of his day, except his enemies, who of course were anxious to be shocked. He was not the only man seeking a corner of the earth for himself in the vast confusion of unknown and unexplored America. But the journey down the Mississippi changed his dream from the feasible to the monstrous.

147

He drifted slowly out of the relatively clear Ohio into the giant stream. There was no crying need to hurry and there was a much greater need to find colonists. Along the banks in little groups stood log cabins. A small bit of land was tilled. He landed and saw the chimneys standing outside the houses, the chinks and crevices plastered with clay, the simple decoration of a crucifix, or the hide of a black bear nailed against the logs, beneath a pair of antlers jutting from the wall. French, English, Germans, Scotch Irish, half breeds of all stocks with the Indians, mingled in the life along the bank. He fell upon christenings and weddings, plantings and cabin building. Fiddles squeaked and the dancing was long and boisterous, or evening prayers were solemn and speech crisp and guarded with the mind standing sentinel over the tongue. In the still of the night Burr thought of these things, and the river got into his blood and reached for his heart cords.

He learned life in those cabins where he put up when the river delayed him. In the morning he saw the fires laid, a buckeye back-log, a hickory fore-stick with chips and small wood between. A clean ash board held the johnny cake set before the fire baking. From the long handled frying pan came most of the food. The tea kettle swung from its wooden lug-pole. A conch shell blown heartily recalled the men from the fields for meals. The dog would howl at the blowing and run out to meet the home comers. Then came the dull clatter of pewter spoons and basins or the hollow rattle of empty wooden trenchers. Spinning he had seen in the east but the dyeing of cloth the dull yellow, gotten from the bark of the white walnut, was new to him. The ink used was made of copperas and oak bark, game was a staple, pumpkin pies were for dessert and somewhere was a bottle of cognac, if the house were French, saved for some high occasion. Everything the people had they made, and much of their heavy work was aided by the nearness of the river. Aaron Burr learned frontier life at first hand and the river was getting in its work.

As he went on toward New Orleans he passed the deeply-laden flat boats. His boat went down stream beside them till they dropped behind with their lazy sweeps. They brought down flour and pork, cider from wild apples, brandy and salt, much iron and a little copper. Great deckloads bulged up out of them. Past him to the northward climbed the keel boats and heavy barges. It was slow work against the current, but they passed and carried up the river sugar and hides. All day Burr's boat travelled among them losing one lot to find another. It was then Burr lost his sense of proportion.

He was caught in the coils of immensity. What was the safety of a little colony in such a valley? Here was something that dwarfed his every idea. Boston, New York, Philadelphia, Washington—journeys between those cities were one thing, but this endless roaming in country that swallowed one was another. He was confused, stunned by the vigor of the life he saw on every hand, the promise of what some day could be. Less and less clearly he saw his rival, the president, the politicians of his old world, the power of the government he had fought to set up in the far removed east. In the days he thought of these things, but in the nights he dreamed.

It was then the east faded from his sight. He saw the Mississippi in flood, an enormous power dividing his old world from his new. Here was a realm of young men ready to be fired to brave action. They cared very little for the ideals of the young nation of which they felt themselves scarcely a part. He had talked with them and knew that in their hearts was the might of the great river, brooking no interference. It needed but a leader and Burr knew himself to be that leader. Every swirling pool, the lapping voice of the river sucking at the cane brake, the full power of its stream carrying down logs and huge roots torn from the far north land, all came to his mind as a symbol of power. Under the dome of heaven he looked upon the points of starlight and felt himself borne away in happy ecstasy. It was his country, this far-flung realm of sturdy works, looking for a lord, seeking a power like the river's which would be theirs, which could guide them and their work. His mind had found an empire that dwarfed the greatest successes of Napoleon, a new world kingdom. In place of ermine there was the fur of the beaver, the feathery host that lifted from the bayous at daylight were fairer than the gilded birds of metal the first consul carried upon his standards into the conquered nations. The court would be a different thing than the effete and vulgar imitation of kingly splendor set up by the Bonaparte in proud display. His dream was rather of the barbaric glory of a far eastern monarch. Feather head-dresses and coon-skin caps would mingle in his followers with Spanish combs and French cloaks. It would be an extravagant, surprising following from all that vast territory and the wealth would be real, not the paper riches of finance, but the creations of his workers wrung from the land itself. He had the greatest highway riches could ask for, the most wonderful piece of unspoiled land ever granted an emperor to fashion to his purpose. He dreamed as deeply as Napoleon and he found opportunity ready at his hand. So, in the nights he meditated, and his vision was stimulated by the country through which he moved.

At Lexington he saw three thousand subjects, at Nashville he thought of the promise of permanence in the eight brick houses out of the hundred twenty that comprised the town. At Natchez he listened to the wild voices undirected in their search for pleasure rising from Natchez under the hill where twenty-five hundred boatmen frolicked as roughly as they worked in the day. Life was not expensive. Beef in Kentucky he found came at three pence a pound, linen at two shillings, six pence a yard. Luxuries were hard to come at, but even they were procurable. A dozen knives and forks cost eighteen shillings and a pair of kid slippers thirteen shillings. The trade of his dream empire needed export of its plenty, dealings with the Indies, with Mexico, with France and Spain. Everywhere he went he felt himself master already, close upon his dream, and farther and farther slipped from view the menace of his old government where he had killed his enemy and lost his political credit.

He had much talk, some of it rather wild. Cabin people were honored by his presence and listened respectfully to him, often catching much of his fire. Young army officers felt him much more a man than their own superiors, fastened as they were within the bonds of the service and loyalty to the government at Washington. Even among boatmen he won friends, which was difficult for they were rovers, turning their hand to things of the moment and looking always toward the next brutal spell of play ashore. He felt sure of the aid of Wilkinson. The farther he left him behind the surer seemed the aid of this unscrupulous commander. Andrew Jackson he felt to be more difficult. He was loyal to the United States, but he hated the Spanish just across the river. A fighter could be urged past his intentions in the heat of battle. Burr felt he could use him. At last he reached New Orleans where he made converts easily. But the honest Claiborne glimpsed his intent, saw the gleam of conquest in his eye and grew suspicious. The Mississippi had fired Aaron Burr's ambition, but to Claiborne it was a symbol of loyalty, always present, always willing to serve, bringing trade to New Orleans, offering a means of sending back despatches to Washington speaking of the enemies of the United States. Burr misunderstood his man, and filled with his vision, turned back to Washington himself.

He lacked funds and the new scale upon which he hoped to operate demanded more money than he could expect to coax from his colonists. In Washington he was all buoyancy. Audaciously he moved under the eyes of his enemies. None wished to share the fate of Alexander Hamilton, but friends were hard to find. Burr had tried the British before, with no success. He was too proud to approach them again, yet he was not too proud to undertake an intrigue with the Spanish ministry. His dream was conquering him, twisting the moral fibre of him all awry. The spell of brooding under the stars was strong in his blood. He needed money and to get it he was not above defrauding the Spaniards, whose destruction he expected to accomplish. The size of his dream, the twisting and turning of his Mississippi blinded him to all else, and his personal honor, to preserve which he had fought Hamilton, he now flung carelessly aside. A man of mystery, intent upon his affairs, he was seen everywhere within the capital. He had lost none of his power to captivate men and women alike. His life was still filled with respect and admiration there in the south, but his vision had done for him even then. His selfishness, his ambition had turned a glorious memory of the world's greatest river into a goad driving him to break faith with his friends to deny his reason the right to stop his headlong rush upon failure, to turn treasonably upon the nation for which he had fought.

By August 1806 he was again on his way west. Drifting down the Ohio he came upon an island where stood a great house, daringly situated, a house of beauty and taste in the midst of an estate. Beside its wonders the humble log cabins of the river folk seemed more than ever pinched and squalid. It spelled romance to Aaron Burr. Perhaps it bore some resemblance to the grandeur he dreamed of as his own when his empire was a reality. He landed and presented himself to the owner. It was the land of Blennerhassett, a rich Irishman who lived in this splendid seclusion with his wife. They were impulsive people who were charmed to entertain a man of such standing. His ready speech won them completely. Living as they did, beyond the usual reach of the government, the United States seemed nothing to them. Impetuously they flung themselves into the adventure, advancing money and undertaking to arouse their neighbors to the opportunities outlined so cleverly by Burr.

Now he saw all as under a shining sun. Journeying down the river he found encouragement after encouragement. In Kentucky, a United States Senator had taken up his case ardently. In Tennessee, Andrew Jackson was eager to seize the power of the state in his hands and call out the militia to invade Texas and Mexico. Burr talked delightedly of the future. If he were not quite frank he was not to blame for that. No rascal

154

can be asked to tell on himself. His words were at work in ways he did not dream of.

From all the river cities where he had touched, had from time to time gone out memoranda, sometimes to state authorities, a few to the president. Claiborne had recognized the danger of the man, added to the demand for separation already so strong along the river, and had done his duty. Rumor attached itself to the soldier of fortune, travelling with him everywhere, Federalists felt they owed it to their party even before their government to check his dangerous ideas. No one could be very certain what he was after, for too much of the matter rested upon mere hearsay, but officials everywhere were suspicious. Henry Clay, then beginning his famous career, acted as counsel for Aaron Burr. He felt it his duty to know where his client stood so he exacted from him an oath of loyalty to the United States before he would represent him. Burr could see only his dream. He was insensible to all else. His oath was given lightly for he would have sworn to anything which would help his schemes. Andrew Jackson, disturbed by the rumors going up and down the river, demanded a similar oath and it was taken with similar nonchalance.

Next, Wilkinson, who had tested out his officers and found they were loyal to their government and would not listen to Burr's plans, turned triple traitor. He was legally in the service of the United States and illegally in the pay of the Spanish. Feeling it needful that Jefferson should be especially grateful to him, so that he might have the president on his side, should his treasonable relations with Spain be discovered, he betrayed Aaron Burr, who had trusted him and been encouraged by him. What was more, he magnified the danger to the nation beyond all limits, so that as its discoverer he placed the government deeply in his debt. Treason, treason, his tongue bellowed the words for all to hear. A cunning rogue was Wilkinson.

Burr knew that he would have no more time to prepare his plan. He had bought land across the Mississippi for forty thousand dollars. Taking Blennerhassett's boats he set off to reach it, thinking it better to be arrested in the south, if he must be taken into custody, than upon the Ohio where his Irish friend had been able to arouse no enthusiasm for the possession of Spanish land. It was a bitter journey, begun untimely in December. He escaped the Ohio militia, but his empire was never to be. He was after all not like the river that bore him southward. His actions were unable to match his words. His vision had undone him. The strength of the great flood was not his strength. Facile, clever, with a flash of daring, he was never stupid. He had planned this journey, but under very different conditions. A mighty argosy was to have moved down boldly. By its size and grand manner it would have inspired all who watched. River men were to have felt its call as a force they would eagerly join. Frontiersmen were expected to leave the cabins they had built with their own hands, and flock to the standard of the new emperor on his way to seize a splendid domain. Instead, only a few followers in fewer boats joined his flotilla, men who were simple minded, honest souls caught by the magnetism of their leader, and who placed that above all sensible consideration or logical thought.

It was flight, that voyage. Every post of the government was turned against the soldier of fortune. Men are always willing to growl, to recite their wrongs, to listen to promises of wealth and power, but at the border of treason normal men hesitate. It was that hesitation which undid Aaron Burr, that and his inability to be as great as the Mississippi which had promised him so much in his brooding vigils along its shores. Even Aaron Burr could not keep the belief of his men at fever heat all up and down its winding stretches.

At Natchez the expedition was shrunken and spiritless. Burr was not the man to fight a losing fight. Hope dictated otherwise. If he fled he might later organize again, and upon the surge of success do better. He avoided finality. At Natchez, he gave up his flotilla without a blow being struck. Without violence, unresisting, he vanished, disguised as a boatman.

It was a fitting garb for his hope. The river had proven too great for him. In a very different way it had checked de Soto and swallowed La Salle. He had not been able to master it, instead it had made him one of its creatures, clad in rough clothes, a man seeking solitude and safety.

He had never thought the arm of the thirteen states, which in his heart he despised, could prove so long. Rejected by the river as its master, he had nothing to hope for from the antagonistic politicians of his home government. He travelled east, his back toward the river, his face set toward the Spanish of Florida, the Spanish whom he had first intended to devour and dispossess in the west, the Spanish he had later tried to secure as financial backers for his undertaking. Hurrying to reach the border before the slow travel of news might outstrip him, he pushed on, still wearing the garb of the river he had lost.

He was arrested by soldiers of the same army he had once officered so gallantly. Brought to Richmond he stood trial. There was no use in telling his judges of the promises the river voices had made him under the starlight. They would not understand. They were men under authority; he felt himself to be authority itself. That had been part of the promise of the river to him. The greatest of our chief justices, John Marshall, presided at the trial of the former vice-president of the United States. Burr still had friends and in that period friends were likely to be true. His father, as president of Princeton College, his grandfather the great Jonathan Edwards, these dead spoke for him, pleading his cause. Even there he could not stand alone; how shrunken was his greatness then. Scorned, but still refusing pity, say that for him gladly, he stood the scathing ordeal of a trial for his dream, and at the end was acquitted on a technicality.

It seems that to condemn him legally as a traitor, it was necessary to prove he had led from the first the expedition that left the island of Blennerhassett. This he had not done, but had joined it later at the mouth of the Ohio River. Perhaps it was a way to save his pride, yet end the danger of him. It of course spelled ruin for Blennerhassett, who had already been ruined anyway. The Ohio militia had shot up his house for the fun of seeing the plaster fly. He had gone south to grow cotton, but he never retrieved his fortune. Wilkinson, the commander-in-chief of the American army turned state's evidence at the trial. He had never been true to either government or man, but in court he was a pitiable figure, for the testimony revealed him as a scoundrel at every turn. Yet Jackson, for some reason or other, left him at the head of the army despite his proven faithlessness. A traitor to his friend, he could not but be despised, but officially he suffered not a wit, but went on to prove his inefficiency and display what a poor soldier he really was upon the northern frontier in the War of 1812.

Aaron Burr was tormented by seeing his bright vision turn into a fiasco, looking vain and foolish indeed when viewed in the light of reason. Richmond was kind to him, but his personal pride was great. He went no more into the west. Taking ship for Europe he sailed for foreign courts. Whether he conspired further there, or merely tried to soften for himself the sting of a ridiculous failure, no one knows. Discredited, he lived a rather unwelcome guest abroad, a man looked upon as foolish rather than dangerous. He saw Napoleon riding on into the heyday of his glory, emperor of half of Europe. Stung by his own moody thoughts, but rid of his delusion cast upon him by the wonder river, he came home to New York. There he practised law for a living for a few years before weighed down by shame, his brilliance gone, life lost its last glamour for him and he died. The great river, however, still flowed to the sea, heeding only its own way, unbridled and unmastered, heedless of his heart-aches. It was still the Mississippi.

✜✜✜✜✜✜✜✜✜✜✜✜✜✜✜✜✜✜✜✜✜✜✜✜✜✜✜✜✜✜✜✜✜✜✜

CHAPTER EIGHT

UNDER STEAM

A CHANNEL for adventure leading certainly upon surprise—that was the Mississippi to the explorers, to the seekers after treasure and the hungry for power. Yet to all that early procession the river's muddy current was a symbol. It carried them in their dreams far from the turgid flow. It served to show their spirit, their bravery, but the river, itself, remained a thing apart. They came never to begin and end in its existence. Their hope was to celebrate the close of their wanderings far from its banks with the river mist and the mud. Triumphs were to be carried away down that same current, and so, across the seas or over the mountains into the heart of the lands of home. If they died on the quest, that was beyond their intentions. Thus cut short, their lives seemed failures. They had not won through and escaped from the savage outpost, the half-known wilderness. They would never sit in cultured seclusion and speak their wisdom roundly. That was for more fortunate men who followed the river's channel home to their rewards. Yet even these time stopped, but the Mississippi flowed on heeding not day, nor night, nor man.

And men came to its valley boldly. At first there were men only, often solitary, or at most two or three in a party. They were seeking a highway to fortune, and the great stream beyond doubt held for them the thrill of danger, yet it was a merry helpmate if a rough master. These men were relieved of their burdens. They no longer had to carry all they owned on their backs. On the upper reaches of the river they built canoes of the birch, paper thin, narrow, cranky craft that paddled easily to a speed of five knots light, or four fully laden. On the lower river there were no birch trees of a size to make their bark serviceable. There the dugout was used. Cedar for the larger, white pine for the smaller. These craft were hollowed out of single trees by hand. Men shaped them up to their individual taste, pointing them out at both ends. The scooping out of the heart of the tree was done by an adze. A man stood upon the log and swung the keen edge smartly, striking between his feet. In a week's hard work and patient chipping, four men could turn out a thin skinned craft nicely moulded and well fitted to the work of the river. Made of one log, she was sure to be tight, unless careless work had started a check in the sides, or the workers, zealous to make her light, had thinned the bottom too much and an ugly knot had worked loose. For ribs to support the sides, natural bulkheads were used by leaving the trunk wood in place between compartments. In a canoe twenty feet long this was done at intervals of four feet, in a thirty footer, at intervals of five feet. The beam was seldom four feet and usually only three, so that the canoes were scarcely good sailing craft, yet they carried mast and yard to spread a square sail when the wind drew astern. By using the various compartments as tanks they were able to load bear oil at the fur stations. They had no other way to carry it, and the traders none other to ship it, so the canoes were filled, a hide served as a tarpaulin, well fastened, and the cargo went down river to market. It was a trick that served for other ladings. Wild honey could be so carried and molasses, where the barrel would have made mean loading and bad windage for the small craft.

The Mississippi, on the whole, used the little craft kindly. Other men had come and gone, but these men and their cargo-laden boats plied their trade methodically. They expected to go home to no king to bask in glory. They had come to the river to make a livelihood. Born out of necessity, their efforts were loyal. They were Mississippi men at heart. The days of long labor at paddle and sail, the nights spent tied to the bank or crossing stealthily the brightness of the moon-track, the swirling tug of the current under them, these were the worship of the faithful who looked for no life apart from the river, who were ready to take their play along its banks. They were a queer lot, not trappers, nor hunters, nor frontiersmen, nor sailors, but a compound of all. As changeable as the Mississippi itself, there was no constancy to their moods, but their determination to stick by the river never wavered. The first of their kind, from them was to come that mixture of good and bad, the typical riverman, a contradiction throughout all his habits, an individual to be

met nowhere, but upon a stream shadowed by French, Spanish and Indian life, yet strong in the surge of its own blood, full of rugged, picturesque life.

Up and down stream the canoes shuttled. Men came out of the east, with their families, seeking transport to new lands. For this the canoes were inadequate. The rivermen had to create new boats, boats able to handle their passenger traffic as well as carry furs. When it came to larger craft they knew only of ship's boats, or packet-like vessels such as the "galley." These were ill fitted to river work, so instead they built the keel-boats.

These were well modelled and double ended. Long, slender and elegant, in their day they generally carried from fifteen to thirty tons. Some reached eighty feet in length and eighteen in beam. They were fine lined, sharp at both ends, drew little water and were lightly built. Clippers of the river, their mission was one of speed and the draft was kept as little as possible that they might take advantage of every favoring eddy, each idling back water. The hold was four feet deep. Its cargo box stood four feet higher. This was sometimes decked in, but more often left open, spanned only from fore to aft by a "gallows frame," over which in bad weather or at night, a tarpaulin could be stretched for the protection of the passengers. At bow and stern were small decks with a windlass or capstan rigged upon each. To these could be led wharf lines or anchor cables. By them the keel-boat could be forced over bars or warped through white water.

Along each side of the cargo boat ran a narrow walk not more than two feet wide. Cleats were nailed across its planks to give better foothold. Ten or more of the crew would range themselves along this walk, each armed with a stout ash-pole tipped with iron. Thrusting its point into the bottom the other end was placed against the shoulder and with the twenty feet of pole fast in the river mud they walked aft as the boat moved forward under their united thrust. When they reached the stern they disengaged the pole from the bottom, leaped to the roof of the cargo box and running forward to the deck reset their poles and again walked the narrow cat-walk. With twenty men at work there were always sixteen poles driving the boat and four returning to begin poling again. Heavy work, hours spent thus in sun or rain, heart, and back, and brain struggling to devour the weary miles and that at the pace of a walking man. Yet this was travel de luxe on the Father of Waters, express journeying only to be enjoyed at preferred cost.

Poling could only be used where the bottom was favorable. In places where the mud was too soft or the current had scooped out deep water under the bank, or where rock ledges made the footing slippery and uncertain, the keel-boats depended upon the cordelle. This was a line about a thousand feet long. The keel-boat had a mast, sometimes thirty feet high. To this the cordelle was made fast and led forward to the bow where a bridle was fastened to it and to the stem of the keel-boat. From there it was taken ashore where the crew "tailed on" dragging the boat against the current. The bridle steadied the boat when the speed was so slow that the rudder was useless. The length of the cordelle lessened the tendency to draw the boat toward the shore. Being fast to the top of the mast it was clear of both snags in the water and brush along the banks. When the wind was fair, sail was set. A square sail on the mast, and at every available angle, blankets, hatch covers, all manner of gear spread to catch every breath. To a deep-water man such trifling seemed childish, but to rivermen, every mile made was a backache less, every hour gained was an hour ashore.

For heavy freight the barge was employed. This was the size of an Atlantic schooner, but with a raised and outlandish looking deck. It carried sails rigged and masted like a sea vessel. Poling scarcely served for its fifty to one hundred tons so handily as for the swift keel-boats, but it was undertaken when head winds and high water made it necessary. In the lower Mississippi a barge was sometimes compelled to resort to warping, the most laborious and the slowest of river methods. Two yawls, one in advance of the other carried out long cables from the barge to a tree where it was made fast. The barge pulled itself up by its capstan to the first tree. The second yawl then had its cable fast and ready, and coiling its used warp the first yawl went above to find a new tree. Six or eight miles a day was good progress. Heavy hauling this at slow speed. A barge required a hundred days to go from New Orleans to Cincinnati. No meaner combination of heavy bodily work and infinite patience has ever been demanded of men. Nothing but brute strength made it possible. Like many other heroic things it grew out of necessity. Pioneers came into the valley, disposed to live on the fertile, virgin land. Their only chance to get things from New York and foreign ports was through the barge and keel-boat. For the going downstream anything would serve.

Any man could build a flatboat. They varied in size and style at the will of the builder. Forty feet over all, ten wide and eight deep, was the size of an average boat of good proportion. No form is so easy to build boldly and badly as this flat-bottomed scow-shaped boat, and many a Mississippi flatboat collapsed at the first hogging strain a misplaced sand-bar and a falling river brought to bear upon her. The middle of these boats was given up to a solid mass of cargo, save for a gangway fore and aft. In the stern was a cabin for crew and passengers. Berths, stools, tables, and a brick hearth for cooking were sure to be found. Since the

167

flatboat had no way of its own in excess of the current that carried it, steering had to be done by pulling its head now in this, now in that direction. Bow sweeps helped out a stern steering oar. As a class the flatboats were clumsy. They were often spoken of, half in contempt, as "broad horns", a term beyond explanation, illogical, yet in some way highly descriptive and a word worth tasting. Broad horn! There it is and it belongs to that ungainly monster straddling the great river, the flatboat.

Flatboats went only one way. When they reached New Orleans they were broken up and the owner sold them in the form of a pile of hand-hewn lumber. Sometimes a few bags of spikes were salvaged, but most boats were put together with wooden pegs like the English tree-nails, which held the old frigate planking to the frames.

Flatboatmen reaped two profits at a leap: their cargoes were sold, then their boats also. This meant that they faced the problem of getting up the river with their wealth. From the days of the Revolution this was difficult, and the growth of the population made it only more so. In New Orleans were many who made it a business to lighten the burdens of the suddenly enriched. Boatmen had first to win free from these allurements, which looked so much like luxury and ease to men who had stood anxiously watch and watch at the sweeps of a flatboat on the passage down. Then in parties they hired canoes or keel-boats up the river to Natchez. There they scattered to the four winds. Some were robbed and murdered by bandits. Marauding Indians still dared attack any caught off guard. The wilderness swallowed some, who became lost among the barrens or mired beyond hope. A few drowned. The man who brought his profits safely home was a lucky fellow. He had done something worth while. New Orleans and Natchez were to him well known, and he could talk of them in his own style. The effort was worth the risk. The reward saw to that.

Another type of flatboatman used his boat to emigrate. He loaded his family and all its possessions, including the live stock, upon a home-built craft. Even the boat he built of lumber he intended later to salvage and turn it into a house to help him gain a foothold on the new site of the home. Occasionally he would have

a brush with the Indians along shore. Either a rise or fall of the river threatened the very life of his children. Snags and rocks lay in wait for him. A frontier fight with friends might find him blinded or lamed. These hazards he bravely faced, moving down the river in his ark, seeking land that pleased his eye, looking for a place to earn a livelihood and raise his family.

Here and there, often about some older trading post or river station, these emigrants hovered for a time and finally settled. So grew up the river cities. St. Louis, New Madrid, Natchez on the Mississippi, and on the Ohio, Louisville and Cincinnati. Politics, world history, only slightly touched the life in these towns. They were too intent upon being fed, fattening whenever they could upon the ever shifting string of homely, squat flatboats, the awkward barges and strutting keel-boats, proud of their relative speed and important because of their passengers. The river people were growing and the midland was on its way out of the wilderness days, but the progress was slow, far too slow for a rising nation. It was then destiny took a hand.

In the effete east, as the westerner thought of it, Fulton and Livingston built a crude hull and set an engine upon it bedded in a stone foundation. Nicholas Roosevelt patented the use of the side paddle wheels and joined the other two. Knowing that eastern people were going west at a surprising rate and in great numbers, these men were quick to see the opportunity of the steamboat in the Mississippi valley. But there was a great drawback. None of them knew what conditions were beyond the Alleghenies. There was only one way to learn. Nicholas Roosevelt went out to see.

At Pittsburg he built a flatboat. It was a floating home for his young bride. The quarters were arranged like a yacht's. A stateroom for the owner, a dining cabin and a pantry were built aft. The crew lived forward beside the galley. From a great stone fireplace the food was sent forward to the boatmen and aft to the Roosevelts. Every comfort that could be looked for was provided. The year was early for daring honeymooners, but Mrs. Roosevelt entered gaily enough into the spirit of the thing and the long jaunt began. The boat was nicely decked affording a flat place where the Roosevelts lived in fine weather. Chairs were provided and a tarpaulin to keep off the sun and light rain, while before their eyes passed an ever changing panorama. This was in May, 1809.

Roosevelt was in no hurry. A nicely modelled row boat was stowed on deck. From this, soundings were taken in advance of the flatboat and the secret of the river bottom stowed away for future service. At any point on the river one stubborn, unavoidable fact might be learned that would knock the idea of western steamboating higher than a kite. Therefore the flatboat travelled leisurely, tying up here for a week, there for three, while its owner went ashore to the pioneers, talked to rivermen, asking for information, trying to excite popular interest in his project. On every hand he met doubt, and at times flat denial that he could put steamboats in the west.

Heads wagged and thighs were slapped in the heat of argument. The boats would not be powerful enough to handle the currents. At low water they would not be able to float. There were points where even the flatboat would not be able to get through. The Falls of the Ohio was one of these, and the passage of them shook slightly Roosevelt's confidence, but he could not confess so much. He listened to all and the talkers were of all sorts. He was told it was not a proper way to go roaring upstream in the face of God's own current; evil was bound to come of such impudence. Any man with half a wit knew that. Suppose the engines broke down; the Indians and bandits would cut the throats of all on board. A flatboat was safer. All this Roosevelt

172

weighed and considered while he went on and found the way open from day to day.

He had started out to see if steam travel was possible. Every mile he voyaged be became more determined to overcome every obstacle. In the beginning he was anxious to find the truth or falsity of an idea. Before he had left the Ohio behind he was fired with determination, caught by the fancy of the thing. From there on he measured the trade the country could supply. He guessed shrewdly as to the probable growth of the population. On either bank he passed villages in the making, settlers hard at it with their axes opening clearings in the wilderness, and he saw a future greater than that of any river in the world. His intentions began to take form. He found coal for his fuel close to the river and arranged his own supply.

The summer wore on. The honeymoon continued. No bridegroom had a better chance to inspire a bride with trust. Indians boarded him and pistols in hand he faced them, armed as they were, gained their respect and tactfully got rid of them by giving them the fire water they desired. Snags and strandings tried the nerves of the three deck hands, the pilot and the cook, not to mention the owner and his wife. River pirates lurked along the shadowed islands, thieves camped upon the shores waiting to way-lay paid off flatboat-men northward bound. The heat increased, growing more sultry, broken by black squalls and thunder. In August they reached Natchez and Roosevelt knew that he wanted to build a steamboat. The Ohio had been interesting but this greater Mississippi, here was a world for his paddle wheels to turn. He saw in his vision settlers carried out with speed and comfort, the land peopled with rugged, full-blooded strength, new towns springing, new trade for new boats.

He talked to his wife of it and suddenly there was great haste on them. The flatboat was slow. Man and wife set out for New Orleans in a pulling boat. When they found hospitality in the tall houses they slept ashore. Four nights of the eight they were afloat, Mrs. Roosevelt, wrapped in a cloak, slept close to her bridegroom who dozed propped in the stern sheets of the boat. They were gorgeous nights passed in the "hotel of the beautiful stars," as the French call such sleeping out of doors. This was the real stuff of romance, a building of confidence and understanding to last them a lifetime.

Once at New Orleans they took ship for New York. There Roosevelt made his report, organized the Ohio Steamboat Navigation Company and ordered his materials sent out to Pittsburg. There he set up his first steamer. His mechanics he brought from New York. The timbers were gotten from the forest and rafted down to the yard. High water caught him and nearly undid all he had accomplished but he persevered. In March 1811 the craft was launched. She was one hundred sixteen feet over all with an extreme beam of twenty feet. Her midship sections were full but she was sharp toward the ends. Two masts and sails, side wheels in a sort of crated paddle box, and a high pilot house were hers. What seemed more miraculous she steered well and engines functioned perfectly. Roosevelt named her the *New Orleans*. Then he turned back to adventure.

Mrs. Roosevelt sailed with him in spite of protests and as they glided down between the wooded hills they listened to the beat of the engines, the flopping of the paddle wheels, the tinkling fountain of the bow jet where the stem piece cut the water. Far into the night they listened, forgetting all the interval between, recalling only the nights of eager hope in the mists of the lower Mississippi. It had been possible in theory. Now they would prove it true beyond all dispute. There would never come another night like that first on the *New Orleans*. It was not a matter for words. It was an affair where solitude was made more perfect by their comradeship and their silence. Grey blue, a steaming shadow, their boat slipped smoothly through the darkness. They would win and open a gigantic venture, or lose and perhaps die if the pilot went adrift. At the flying speed of nine knots with the whole thrust of the river behind them the damage of a stranding would be great. Even drifting flatboats went on so hard they sometimes broke their backs. Who could tell how the *New Orleans* might pile up? But the Roosevelts were confident; they would not think upon the chances. Instead they smiled at each other and joyed in the throb of the engines, the first engines the valley had ever known.

They made Louisville on October first at midnight.
Letting go the anchor they hauled round, head up-
stream, and blew off their boilers. All the town thought
a floating sawmill had come down, or the last day was
due and the world was sizzling under the doom heat.
When they learned better they thought something
should be done about such an arrival so they gave a
dinner to the boat's people and Mr. Roosevelt countered with one in the after cabin of the *New Orleans*.
Meanwhile he had found a profound skepticism
among the townsmen. They came to his dinner. When
they were well settled suddenly the deck beneath their
feet shook with vibration, the engines turned over, the
New Orleans was off. The guests tried to look like
guests, but it was no use. They left the tables in a body.
There was a reason for their fear. Not only did they
mistrust the behavior of this queer monster to which
they had been decoyed, but below the city waited the
Falls of the Ohio. Now the great doubt they had held
was that this fool affair would ever be able to buck the
sweep of the Ohio. Roosevelt had heard it everywhere
ashore. He met it in taunt, in general talk, in jest. His
diners were going to have the demonstration of their
folly at the expense of their fears. It was worth the
victuals. He headed the *New Orleans* up-stream and
she steamed up into it under a full head of pressure,
throwing sheets of spray to either side, storming along
in fine style. Louisville dropped behind; the boat could
travel up-stream and that at a good rate. Roosevelt
killed prejudice at a single stroke, and sent his converts

176

home happy over a good dinner to boot. Then he'
climbed the river to Cincinnati, that all along the river
might see the steamboat was practical. She churned
along, sounding her whistle at every landing. Roosevelt was determined to make the pioneers conscious that
here was a thing come into their country that defied the
primitive conditions which bound them so straitly,
that went about its business heedless of weather and
low water, able to extricate itself from any predica-
ment into which it fell. On his return to Louisville the
river was high and he was able to go on.

The Falls of the Ohio were not a real falls, but rather
a rapids where the water flowed over a rocky bottom.
Falls or no falls it was a ticklish passage for the *New
Orleans* to attempt. They were determined not to turn
back. The engineer fired up. The smoke broke from
her funnels in clouds. The owners came on deck, Mrs.
Roosevelt a little pale, her husband grimly quiet. They
waited for the pilot's bell to ring, for the first kick of
the engines that would drive them either to destruction
or to the open water that was not stopped again before it
poured into the Mississippi. All six deck hands stood
by the rail affecting easy banter, but it was a half
hearted attempt. Every soul on the boat felt the strain
and knew the risks. There were places down the wide
shute ahead where the water was but half a foot deeper
than the boat. The bell sounded. Gently the *New
Orleans* moved out into the stream. Steam curled from
under the safety valve, hissing loudly. The pilot twirled
his wheel, felt the inadequacy of his rudder with his

boat moving so slowly in the strengthening current and
rang the jingle for full speed. Passengers gaped, as
passengers always will. Down through the white water
they whirled, all in the pilot's hands. The destiny of the
first steamboat in the west rested upon his knowledge.
Almost before those on board knew it the passage was
made and the steamer stopped to drop the fall's pilot
beside Sand Island. The six deck hands pummelled
each other merrily, the engineer wiped his face and

178

told his firemen to spare the wood. The Roosevelts looked back, where the water whirled past the point of the island and wondered at their courage. Ahead lay flattened land and peaceful water. There was no sign of need for distrust. Then in a moment it began.

The water was broken into senseless waves like those in a teacup when a violin bow is drawn across its rim. The river banks lifted and fell, their movement horribly visible to those on board the steamboat. The hull shook until everything loose rattled. Nausea came upon the voyagers like seasickness, only swifter. Flesh and blood could not withstand the shock of seeing the world reel and stagger. Along shore a bluff fell clattering into the river. The anchor cable of the *New Orleans* tautened like a whip lash. Nicholas Roosevelt ordered the boat forward at once. He might be caught in an earthquake, but he did not intend to lose his passengers, so he got underway before they could collect their wits enough to desire to be left behind.

The earthquake was the great New Madrid disturbance, the most far-reaching the Mississippi valley has known within the days of man. It had been a bad year, that of 1811. The comet had frightened the people almost to death and this new catastrophe rumbled upon them out of the earth's core. But the *New Orleans* pushed on. Her fuel now was coal brought from upriver for this long stretch of little known waterway. It was well she carried it, for the earthquake, which was felt in new shocks from time to time, had scarcely left a land mark the pilot could recognize. There was no landing. Islands had sunk or were in new shapes. Dead trees, blanched and white for years, had disappeared in a minute, tumbled from a crowning bluff, or, if in a bottom, perhaps sunk beneath the flood, for the water was high at the time of the earthquake. Progress had to be slow on such a river. It was as if no one had ever gone down it before them. The leadsmen worked night and day as the pilot felt his way along. The slowness of this progress burnt up their fuel faster than they expected, so when the coal was gone they had at last to stop to cut wood which was green and made steam very slowly. Harassed, continually worried, all on board suffered nervously. Occasional shocks did not make their minds any easier, nor the stories they picked up at every landing.

It had been a mean experience on board the *New Orleans,* but the vibration of the engines and the natural sense of motion had lessened the effect of the rocking earth. On shore the earthquake had done great damage. As they worked down-stream they came upon miracle after miracle. Men came to them and told of low brush rising and falling in waves, of a mighty whisper of leaves tossed and twisted by the shocks. All agreed that the beginning came upon a windless day, yet trees were lashed back and forth so savagely limbs were torn from them. The earth's surface broke into great gas bubbles, solid ground turning to a thin crust through which a fleeing man stepped as he might have through rotten ice. All that happened in the woods and fields had its counterpart in action under the river bed.

Geysers broke through the water, spouting gas, riling the water to an unholy red. On the shelving shore of an island, craters opened explosively, leaving a pit sixty feet across and a fourth as deep. By New Madrid, great fissures had opened, all leading from northwest to southeast. Men spent days astride logs they had felled across these fissures to keep themselves from being engulfed. There were many in the town who begged to be taken on board the *New Orleans,* but she had food only for her own passengers and was bent first upon her own safety. A sleek, blue craft, she passed on down the river through the lessening aftershocks, riding a current red as blood save where it was broken by a yellow rile of bashed reef, or shifting sand bar newly stirred by the earthquake.

It was a long, hard run from New Madrid to the southward. There were no towns to furnish them with help let alone undertake repairs on this first of her kind to come down river. The Mississippi has always been shortening itself by cutting off headlands, making first an island of a cape and then, once the new passage was cut, washing down the island at the rate sometimes of many acres in a single night. The earthquake had added to this natural process of change a perplexing uncertainty. To the pilot all was strange water and a new river. At the north of the White River they found a settlement which had no horror tales to tell of the earthquake. This was cheering news. In the lower river they could hope for more usual conditions. It dawned upon Nicholas Roosevelt that his boat had been called on to meet its test at the very heart of the disturbance. She had born the brunt of it and lived. With good fortune and care he could expect to win through. The black terror of nights spent listening to the sliding rush of falling banks, the complaining of of wrung and tortured trees, and the bellow of some geyser turning loose its shower of mud, that was over, for just ahead lay Natchez.

High on a bluff of the east bank it greeted the tired eyes of the crew. They knew what waited for them in the lower town where the boats lay so thick. Shore leave and pay and a bit of sport! These they might surely hope for after a voyage in which they had experienced everything but a sight of the subterranean fires of whose existence they had seen sufficient proof to satisfy them for a lifetime. Pride sent them ashore gaily. Were they not men of the first steamboat to come down river? Not a man would have changed berths with any of all the crowd. The *New Orleans* had brought them safely through everything. They wanted to finish the passage, but first they wanted Natchez.

The townfolk were eager to see the queer contraption tie up. They lined the wharf, waiting expectantly. When they heard of her passage respect for her grew apace and the town honored the crew. After his entertainment on shore Roosevelt was given very good proof of the faith he had aroused. The *New Orleans* was entrusted with her first consignment of cargo, a lading of cotton. There was the usual shaking of heads and prophecies of woe, but the Roosevelts were used to such things, in fact they no longer heeded them. Instead they cast off and resolutely threshed their way down-river. They were really in the south now. The river fowl wheeled in extended line above them. Beneath they felt the thrust of their engines and saw the greyish water break into white at the bow. A narrow passage, an island set thickly with young cottonwoods, a plantation with a great sweep of land, then the broadened lower river. Keel-boatmen bound up-river hailed lustily. Broad horns, drifting with the current, jeered at them on passing, but when the tall stacks were further off and the gap between astonished them, the flatboatmen leaned on their sweeps and gave the matter long thought. They unknowingly looked upon the death of their trade. That ghostly wisp of vanishing smoke, that tireless catch and spatter of paddle wheels, travelling up-river to him on the wind, were both emblems of his own passing had he but had the wit to see it; but he was impressed only by one thing, the speed of the rival craft. The *New Orleans* had passed him at six knots.

In time the Roosevelts carried their cotton safely to New Orleans. January 12, 1812, was the day. All the wonder of the undertaking came to the old town. Before the adventurers lay almost a crescent of shipping: flatboats, keel-boats, river craft of every sort. The air was clean above them and the sunlight good. Sloops, schooners, brigs, waited at the wharves, ready to carry river cargoes beyond the great delta mouth to all the world. The houses of the Faubourg St. Mary, the sea-going section of New Orleans, were massively built, stuccoed in white or yellow. A French or Spanish town rather than an American. It was not new to the Roosevelts. It was there that his great idea of carrying steam to the west had taken active form. Had he not whiled away days in sight of the cathedral, waiting for the steamer? He knew its four towers. He had heard its two bells at work and gone in to find himself shut off from the rattle of carriages and the hustle of life along the Levee Street. The dead had slept beneath his feet. Images of death, of the invisible world, of eternity, had caught his eye under the yellow light of waxen tapers. His heart had cried out against the shortness of life and the difficulty of doing anything with it, and yet he was so soon back. His hand sought the whistle cord. He must let this city of old shadows know his dream had come true. He had done it. With freight on his decks, his boat sound and able under him, he was ready to go into trade so soon as he could land. He slid down toward the dock. A bell slowed his craft, another stopped her engines. She glided on along the dock wall in a long swoop. Her lines were heaved and her hawsers hauled to the wharf smartly. Still she coasted on. A cry went up from the roustabouts. She was going too fast. She would part her lines. The spliced eyes were dropped over the snubbing posts on the levee. The roustabouts ran. No man wants to have his head broken by the flying end of a parted line. Then the *New Orleans* reversed her engines. Her paddles turned slowly at first, then lashed the water. A white current raced along her side. She stopped in her tracks. Her pilot had not brought her all those miles for nothing. He had learned his boat. The roustabouts, shamefaced, came back to take her springs and breast lines. She was at rest. A thin breath of steam hissed from beneath her safety valve. A lifting stain of smoke floated north from her funnels. The first steamboat had travelled the Father of Waters. The faith of the Roosevelts was justified.

He contracted to sail the *New Orleans* weekly up to Natchez. She was too deep of draught for the upper reaches. For that he would build others, better boats, shallow bottoms, larger overhangs to make loadings easily. This boat which he had, this first of her kind, would stay by the lower river. He would make a passenger rate of twenty-five dollars going north and eighteen coming south. She could breast the current at four miles an hour and even a keel-boat frequently gained only one. Her speed would give her the trade. Roosevelt had but to maintain her, to see she was well piloted and her engines kept fit, and for the rest his profits were sure.

As for the *New Orleans,* she ran gallantly in season and out. More and more cotton, tobacco and hemp was trusted to her. She become a byword to passengers. She lived to carry down General Coffee and his Kentucky riflemen to take part in the defense of New Orleans. She died, when the river, careless of her gallantry and faithfulness, fell suddenly one night. She was tied up, and the falling level lowered her upon a stump, and then, dropping still lower, forced the snag through her bottom by resting all her weight upon her planks until they were broken through. Her crew, confused in the dark could make little of her trouble, and she filled and sank, leaving them just time to make the bank and take their necks out of danger; but the *New Orleans* was done.

✠✠✠✠✠✠✠✠✠✠✠✠✠✠✠✠✠✠✠✠✠✠✠✠✠✠✠✠✠✠✠✠

CHAPTER NINE

TO ARMS FOR THE MISSISSIPPI!

NAPOLEON sold Louisiana to the United States to get money for his wars and to keep the Mississippi from falling into British hands. By 1814 he had spent his money, lost France and found himself Emperor, in name only, of the little island of Elba. If he knew the English plans to invade Louisiana, as he might easily have done, he must have doubted the power of the United States to hold it. With a shrug he may have thought of this effort to thwart England as another mistake, another failure. Our war with England had not been successful. From 1812 on there had been a few heroic frigates simulating the efforts of a navy against the thousand English sails. No great general had risen to command on shore. Our capital, Washington, had been captured and burned. Save for the Great Lakes, no theatre of the war had held success for us. Then to crown their success, the English directed their efforts against the Mississippi. Yes, if Napoleon knew, he must have smiled a little sadly, and thought his trust in the United States as a rival to England misplaced.

The British gave long thought to the matter. A people of empire, used to colonies, their agents were able to assure them that the people of the Mississippi

187

were but half-heartedly American. They pointed to the Creoles who had lived for twelve years, in the face of many hardships, under a flag they felt to be alien. New Orleans was largely peopled by these folk. New Madrid was still strongly Spanish. Never was there such a chance to gain so much with such ease. With the Mississippi in British hands the thirteen states would find themselves hemmed in upon the north by Canada and upon the west and south new territories would be speedily developed as English colonies. So girdled, the states might be willing to return to the empire as a colony. It was well worth a thrust; only the size of the country, the baffling problems of area and distance, threatened the attempt. The Mississippi was not a French nor an English river. Even though conditions should prove favorable many men would be needed, a vast expedition.

The English knew how to colonize. They had studied the climate of the country, accordingly they chose winter for their attempt. No noisome heat in fetid cane brakes and swamps was to thwart these veterans. They might lose men in action, but no more by disease than was unavoidable. Soundly organized, the overseas force gathered in the south of Ireland in 1814 waiting for early winter.

Fifty vessels and nearly twenty thousand men were concentrated. The brother-in-law of the Duke of Wellington, Sir Edward Pakenham, was given command. A thousand guns were to sail against the river mouth, a thousand guns manned by trained crews who had been under fire. The flagship of the admiral, Sir Alexander Cochrane, was a symbol of British prowess. She was the *Tonnant* of eighty guns. Nelson had taken her from the French at the Battle of the Nile. Five "seventy-fours" sailed, all under command of picked men. Sir Thomas Hardy, in whose arms Nelson died at Trafalgar, was of these. Twenty smaller ships, armed with from fifteen to sixteen guns each, stood ready to convoy the sixteen transports. Four regiments of veterans, fresh from the campaigns of the Peninsular War were shipped. Thousands of other troops, none of them green, for Europe had been war torn so long that all soldiery saw action aplenty, were added. A short campaign, overwhelming force, Louisiana British, the upstart states punished; there was a plan that seemed sound. That success might find nothing wanting, civil officers were provided, ready to set up a true British administration when the Mississippi became formally a part of the empire. Napoleon could be excused for doubting the United States under such conditions. These English had locked him out of his own Paris, why should they not take New Orleans?

The armada entered the Gulf of Mexico early in the winter and reached the anchorage at Ship Island, December 10, 1814. Local aid was secured, Spanish fishermen, renegade coast sailors, all able pilots, served the invaders. New Orleans was to be attacked upon the flank by way of Lakes Borgne and Pontchartrain. Able coast captains so recommended. The city still possessed ramparts, but was undefended by guns. Moreover, the citizenry was apathetic. The few Americans mingled among the Creoles could stir no loyalty. The British were reassurred, but there was no careless overconfidence. Veteran blood was cool and well disciplined. The officers were suspicious of the attractive coast, refusing to be lulled into security. Every detail was considered before a blow was struck.

It was a day for a man and Andrew Jackson was the man for the day. He was a Scotch-Irishman born to a Carolina frontier family. Frankly child-like, he was fearless, but ignorant and prejudiced to boot, a quiet man possessing both good and bad in abundance. Stories attached themselves to him. He believed no Indian good but a dead one. It was said that as a boy he followed a deer for a shot one day, when, as he settled to fire, he saw an Indian also following the deer. By the brave's carelessness he believed he had not been seen. He had but one barrel to shoot. If he killed the deer the Indian might shoot him before he could reload. If he fired at the Indian the deer would escape. He thought quickly. Then he swung his muzzle toward the Indian and waited. When the Indian fired at the deer, Jackson fired at the Indian. Each hit his mark and Jackson got the deer. It was typical of the man to meet two problems, each colored by his desire, with one solution. He was a bundle of contrasts. Chivalrous to women, he had only the bullet and bayonet for the Spaniard. He held the Federalists in little more favor. His friendship was difficult to maintain. In spite of the esteem in which he held Senator Benton, he backed a man who picked a quarrel with Benton's brother, standing as a second in the duel between them. This was followed by a wild fracas in an inn where he faced the brothers in a bloody fight. The younger Benton was badly cut and Jackson received two bullet wounds. He was too sturdy to be thus brusquely stopped. Aaron Burr involved him in his scheme, yet at Burr's trial, he took the dreamer's part, delivering a heated speech against his enemies and an open attack upon Jefferson. Andrew Jackson had courage; he did not shrink before the arrival of the British force. His will was indomitable, and he had decided America should not lose the Mississippi, not if the whole British army and navy faced him. He was the man for the moment.

In November he attacked and captured the British post at Pensacola. It took him a week to storm the position. Pensacola was not his first glimpse of real soldiering, he had served in the Revolution, but it was his first taste of active command. There he saw European methods of warfare, so different from his own Indian fighting where the enemy was often little more than a dusky, gliding shadow and in every shadow the sting of failure and sudden death. In his own way he decided upon his tactics. Frontiersmen were used to the rifle. When they fired they shot at a chosen mark. These soldiers of England were automatons. Memories of their fighting in the War of Revolution returned to him, cannon volleys fired at command, battle in formation, war according to regulations and depending upon the brains of the officers rather than the ability of the men.

After Pensacola he fell ill. At first listless, he went down, a very sick man indeed, but there was no time to rest. The danger was now too apparent. Rumors of the size of the British expedition reached him. Every day told of their effecting some planned attack.

Pakenham had been well schooled. It was he who broke the centre of Marmont at Salamanca. Against Napoleon's marshals, Soult, Massena and Victor he had learned warfare at first hand. In his plan of taking the Mississippi he proceeded systematically. No enemies were to be left at his rear or flank. The American gunboats, often but barges with single cannon, were swept from Lake Borgne. Spanish fishermen guided his first movement and a heavy force landed at the Isle aux Poix in the southeastern tip of Louisiana. By lake and bayou, through the marshy country, these troops travelled toward the Mississippi. Six miles from the city they struck the river. The Americans tried to check them. River craft attacked, but by the use of red hot shot they were either sunk or driven off. All through November and December they made more perfect Pakenham's plan, who, having such resources at his hands, determined to leave unconsidered no dangerous factor of the situation. Large as his force was he was nearly four thousand miles from reserves, a whole sea away from London. It behooved him to be cautious.

He heartily despised these backwoodsmen, ragged and unkempt, but he never left them out of consideration. On New Year's day, 1815, his men were safe upon the plain of Chalmette. Re-inforced to eight thousand, his left flank protected by the river, his right by the efforts of his outlying forces, he was ready to attempt the city so near at hand.

But Jackson had recovered. Reports of the British advance reached him, honest reports which let him know the real size of his enemy. When they began to come in he was indifferent, listless after his illness, but as the seriousness of the effort came home to him he roused. Here was a fight worth winning. The great armament progressed methodically. Would they take from him and his Tennessee men the river that meant so much to them? All his young manhood had been passed beside it. He had seen it in flood and drought, but he remembered it then only in flood, whirling down trees, crashing all in its path into a wild jumble of wrack and ruin. The spirit of it touched his blood. It was his river. Its riot of strength was his. He would stem these British no matter what their force. He would come upon them like the river that they should not ascend toward his own hill girt country. At Pensacola they had given in before his storming. This time he would stand firmly in their path and turn them back. His reports were accurate enough to show that the objective of the British leaders was New Orleans. Common sense could have told him as much, but it was better to know beyond all doubt.

Hastening through the wilderness he reached New Orleans on December 2. Nothing had been done. The town was quiet and easy going. Their eyes opened wide when the martial stir of Jackson and his men reverberated from the old walls. First he proclaimed the city under military law. Next he recruited the men.

It was a queer army. A battalion of negro troops was formed. Spaniards and Creoles were mustered, where possible under some old soldier of Napoleon whose hate of the British was perfect. No need to arouse those men. They were already alight with the fires of older battlefields. The easy-going inhabitants prepared like true patriots, drilling well. They had not been disloyal, only disinterested, and that was all gone now, and in its place was a light-hearted will to do that which promised well for the city. Outlaws and desperados there were aplenty in New Orleans, for the city was a place where men minded their own business rather well. Young countries are not particular about the history of their citizens in their need of men. These men were with Jackson to the hilt. There was no joy in them at the prospect of the British administration which might ask embarrassing questions. They were not desirable characters, but they could fight and Jackson was not afraid to handle them.

Perhaps the most curious wing of the motley army was the freebooters led by Jean and Pierre Lafitte. Near the river mouth at Barataria they had a pirate stronghold of long standing. The British had tried to buy them, but they would have none of it and rallied to Jackson's side. Well known, and dreaded as being wholly lawless, there was no doubt of their effectiveness as fighters. They were valuable rascals to have in a pinch since they knew the river and the details of the coast outside for a hundred miles in either direction.

In his heart Jackson believed most in his Kentucky and Tennessee riflemen. Their arms were crude. The barrels of their weapons were made of soft iron. The hooked stocks were ungainly to modern eyes, but the men who shot those rifles knew what they would do to the fraction of an inch. They were used to killing game from squirrels to bison, and most of their shooting was therefore at moving objects. Their lives were always their own in the woods. Every day threatened an Indian fight, where every man took cover, or smirched himself with paint to be nearly invisible in the woods. Stealthy, cautious work that, demanding cold courage and steady nerves. Such men would have despised uniforms had any been available. They could fight under any conditions, but each man thought for himself and found himself the equipment that pleased him most. There were riflemen there who used a tomahawk as skilfully as an Indian and whose knives had taken scalps with a very successful finesse. Under Coffee and Carroll they marched down from Baton Rouge where they had been in camp. Jackson had ordered them not to sleep until they had joined him. Knowing Jackson, they appreciated that there must indeed be an emergency if he felt such orders necessary. Tirelessly they swung down, for if there was to be a battle they had no wish to miss it. They came into the city very timely.

Andrew Jackson had not wasted a moment there. He was determined to keep New Orleans although he had only his men to do it with. There was some little cotton to be had, only a few bales. That he built into his entrenchments. For the rest he used anything he could lay his hands upon. Rails, earth, baulks of timber: out of these he built his gun placements, for he had a few cannon, all too few to front the veteran troops that were sure to come against him. Then, after a final inspection, in a day or two less than a month he mustered his army. Creoles, Spaniards, negroes and pirates, manned his defenses. In the centre of his line there lay a swamp. He could not trust the improvised fighters there. They were not used to fighting knee-deep in water and mud. His woodsmen were better for that ground, his woodsmen and their rifles. They were not used to shooting into masses of troops, but they could be counted on to draw a fine sight, to knick a squirrel's head, or take a bison or bear just behind the shoulders, stopping him with little running. Knee-deep in swamp and unprotected by entrenchment, which it was impossible to dig, they waited patiently. As hunters they were familiar with waiting. Andrew Jackson knew them to the core; it was fighting after their own hearts, to which the English regulars brought no advantage save numbers.

Pakenham found no need to attempt flank attacks. He was ready to assault the position as he found it. He

HE WAS DETERMINED TO KEEP NEW ORLEANS

saw no great danger in the ragged and nondescript rabble that manned the defenses. Jackson had used an old mill-race to check the British advance. Behind it he had mounted what cannon he had. They were few but valiantly served. The only part of the line directly assailable was that in the cypress swamp. The British commander decided to try his cannon fire upon the position, which opened in a roar, right merrily. It shook the Creoles and Spaniards somewhat, but did little damage. The Americans waited and watched. If Pakenham thought to rout Jackson's army by alarm he was sorely mistaken. When they found the cannon thundered out of all proportion to the damage they did, the negroes began to banter, making jest of the fire. The old soldiers of Napoleon steadied the line in their own sections, heartening all by their disregard of the artillery. The backwoodsmen played Indians, taking shelter wisely and for most part escaped unharmed, and even gained some rest. New Orleans could not be captured by noise. It refused to be impressed. Moving big guns was laborious and often impossible, without roads, and over the marshy lands. Pakenham at last desisted.

On January 8, 1815, he decided the time was ripe to end the matter. Trusting that his cannonade had shaken the assurance of the defenders he made a feint before his real advance. Across the river, upon the west bank there was an American battery which commanded the approach to the city. Flinging his attack against that, Pakenham waited for a time and when he thought the reduction of the battery had sufficiently diverted American attention he opened a frontal attack upon the centre of the line. Andrew Jackson had outguessed him. He had made no effort to support the attacked battery far out upon his right, and he had placed his Kentucky and Tennessee men directly in the path of the main attack. Pakenham despised all the colonials and after weeks of patience and caution hurled his whole force into an impetuous storming, which was scarcely sensible or soldierly. The American guns opened at once. All the fire of the defenses poured upon the two columns advancing stalwartly. In close formation the veterans of the Peninsular War came on. Sixty abreast, in solid front, their boldness looked like madness to the American riflemen waiting knee-deep among the cypress roots. The British artillery fire was high, sweeping the tree tops in terrific crashings. The riflemen saw nothing to such work. Not a man of that brave first rank coming on so smoothly could live. Every man knew that he could not miss such marks. There was no dodging from tree to tree, from waving brush to guarding grass tuft. The English never faltered. They fired volleys occasionally, the wooden unnatural gesture of the European soldiers. Few men aimed their guns; it was collective fire supposed to crush by the sweeping weight of the bullets thrown across the space and into the body of the enemy. But there was no body to this enemy, they were individuals well hidden, waiting only for the word to go into action.

When the belt buckles gleamed and the shoulder

IN CLOSE FORMATION THE VETERANS CAME ON

straps of the scarlet coats stood out clearly, the riflemen gauged the distance and played their part. They fired in irregular style, according to the speed of loading. The crackle of their rifles was drowned by the roar of a volley from the enemy, but when the leaves were finished shaking and the British ramrods were clinking in their barrels, busy at the loading, it droned on in a melting, withering fire. Aiming at the belt buckles, the Americans stretched man after man. Some went down cursing, some praying, others with a look of mild surprise. The fire kept up. Pakenham's second in command was killed. The backwoodsmen were not heartless. They pitied their enemies, but they were not going to lose their Mississippi. It was theirs. Every bullet cried that, and the English, stricken dumb by surprise at their reception, faltered and then fell back.

Doubting his leaders, or beside himself at the stubbornness of this hated rabble that dared slay his troops, Pakenham took command of the second assault. If the first had been bad, this was worse. Major-general Gibbs and Keane were shot down. Seven colonels were slain, seventy-five lower officers, and soldiers by the thousand. The English failed to reach their enemy. They had no opportunity to try their bayonets. Even there the Americans might have held them, for backwoodsmen are good rough and tumble fighters and they were not harrassed by bulky equipment and ungainly uniforms. The English would however have greatly outnumbered them. Perhaps it was as well the rifles of Jackson were like the Spirit of Death itself. In the midst of the confusion Pakenham fell, shot through the head. The veterans were withdrawn. The vicious crackling of rifle fire died away, there was no need for formality. The battle was over.

Ten days the British considered conditions, weighing odds, burying their dead, tending their wounded. Then solemnly and silently they began their retreat to their ships. Behind them they left a city wildly hilarious, intent upon celebration. The greatest expedition of the war had been soundly thrashed and sent home minus its three ranking officers. The Mississippi was safe and the western people to whom it was at once willing servant and tyrannous master had done the thing with little help from the east. Louisiana and Tennessee had drawn together to fight for a common cause. They had their river. Together they had confidence. The Mississippi lands were ready to assert themselves in national affairs, they were no longer to leave the government of their lives so completely in the hands of the far off east. They had their own hero, Andrew Jackson, who was soon to be the idol of the whole American people. He was a link in the quickened American sense of nationality, for the battle of New Orleans reminded the east that after all these fighters were of their own, part and parcel of the American population. The Mississippi moved from the dim borderland of unknown districts into the very heart of American possessions. It had been fought for, it had been won. About it was the merit of a reward given as a trust. The nation could never let it go again, not for fear, nor necessity, nor political differences of parties. A nation dare not deny its just pride nor discredit the actions of its great, and Andrew Jackson had saved New Orleans.

In time the British fleet reached home with its tale of catastrophe and its evidence of heavy loss in the gashed ranks, as the survivors of the veteran regiments marched ashore. The news went up into the court, into the houses of government and, in time, word of the defeat of the cream of the English soldiery by a ragged lot of rascals, hidden for most part in brush, and behind entrenchments, sifted down to the people of the nation. Tongues wagged and heads nodded. Time passed and word of the battle crossed the channel into France and so to all Europe. At every step the repute of the Americans waxed greater and greater. Representatives of the United States were treated with a new respect. The Mississippi had bred Andrew Jackson. He was its child, the eldest of all the great Americans it was to bear. From its son had come in a single battle more prestige for his country among nations than the states had ever known. American valor was no longer a provincial, unknown thing. Even Napoleon, lost forever on the English Saint Helena, could have honored it, even he, valor's son, might have been proud to see its like in the French hosts with which he had swept Europe. It was a certainty, properly appreciated; like the powerful reality of the Mississippi itself.

✠✠✠✠✠✠✠✠✠✠✠✠✠✠✠✠✠✠✠✠✠✠✠✠✠✠✠✠✠✠✠✠✠

CHAPTER TEN

PACKETS AND POCKETS

THERE was no light at the head of the short, low pier where the island people waited for the night boat. Down-river, the churning beat of a paddle wheel echoed under the bluff. It ran along the shore, rumbling in the dark, and behind whispered the swing and swish of the wake wave, breaking along the short, clay beach. The blindness of the cove was already filled with the clamor of tumbling water, and the hiss of steam, although the boat was still far down on the edge of an eddy, just out of grips with the Mississippi current. Those on the pier could make out nothing in the obscurity, not the water ten feet from the wharf, nor the hunched shoulder of the point beyond the cove, nor faintest gleam of starlight from the heavens, that seem so high and remote behind the invisible bluff which overhung the landing.

They knew where they were, but the pilot had to pick his way in past a big cottonwood on the lower point. He headed for a low white shed he could not see, but knew was just over his bow. Sagaciously he rounded a "planter," a tree piled up by the current with its arms fast in the mud of the bottom. He knew the spot by some dim twisting of the shadows, some difference in the blackness discerned only in the wisdom of a pilot. Just there in his path had been a "sawyer" a month before. A sawyer was a snag that bobbed up and down under the thrust of the current. Now its black bole would rise, then the water would swirl and eddy over it. There was no seeing on such a night. The pilot chanced it and held in, hoping it had moved down-stream.

On the dock they could make out the steamer's red and green side-lights, hung high from her funnel tops. There were no other lights on her forward, not a gleaming chink. Tarpaulins covered skylights, and the cabin windows were screened to let the pilot use his eyes to the full. There was not even the red glow of pipe or cigar on deck. Even so tiny a fire could blind the pilot's eyes and rob him of his cat-like sight. A hazy gleam shone upon the water from the windows along the side of the boat. The side-lights were closer, high above the pier, like unkempt and discolored stars. Suddenly, out of the darkness, roared the whistle, sounding a thousand times too loud for the cove. In the pause, after its blowing, a man's voice hailed the shadows gustily, from the shore, to assure the boat she was to make a landing. The hail rose thinly to the pilot's house, a curious, substantial, living thing out of the uncertainty of the night. The pilot tapped his big bell three times. Below decks the engineers heard the triple warning to stand by the engines. Soon the pilot pulled a wire to the engine room. The boat slowed to the single stroke upon the engine gong. She slid into the shadow of the land in a way that made the passengers gasp. They felt as if their boat had suddenly been engulfed by the bank, yet she floated clear.

Then, at a word, two streaks of fire moved forward along the boat's deck, one on either side. Men were carrying up brands they had lighted from the fires under the boilers. At the bow they halted for a moment. Two larger flares sprang up, steadied into a glow, and burned brightly. Two iron baskets a foot across and two deep, filled with split southern pine and swung between the prongs of iron forks leaning far out over the water, burned happily, pushing back the darkness. The fat glow of their flare smoked. "Jacks" they were called along the river. Drops of resin caught fire and dropping in the water were carried off by the current, still burning as they floated down-stream. Boys with smutted faces tended the jacks, white boys with black faces. They worked earnestly with evident joy. The orchestra struck up. "I've left my girl in New Orleans" or "Ducks play cards and chickens drink wine." Lines were passed from roustabouts afloat to others ashore. Some freight and baggage was left behind. Words of caution, words of farewell, jests between boatmen and shoremen, the laughter of women, the prattle of children sounded, lifting into the quivering flood of light. Long men in tall hats came on board with dignity, wrapped in cloaks; ladies in billowy folds of skirts, children, small copies of their parents in both dress and manners: all these seen for a moment before the jacks died down. Then the bustle of departure, the knocking of sizzling fire brands from the baskets into the water,

the tang of wood smoke and the scalded embers, the shuffling of deck freight away from the bow.

In a moment the boat was adrift in the dark again. The pilot's sense was at work once more using every memory he had, conscious or unconscious, feeling his way from sand-bar to island, past snags to a chute, crossing the river in the teeth of a log-burdened current to drop four pieces of freight, facing a squall with Arkansas lightning and thunder, to get back to a town where waited two passengers. All the night through that pall of darkness was upon his eyes, the thrust of a laboring engine beneath his feet and upon his heart the responsibility for every life on board. An accident would find the removal of passengers doubtful or even impossible. Such was the normal condition of the river that any mishap might have the most serious of consequences, but the pilot carried on trusting his knowledge and his experience of the river.

That was a Mississippi steamboat. The press hailed them as miracles: master of the waters, conquerors of the hazards of travel, the great controllers of both time and accident. Yet they crept humbly enough in the dark. In fact, at low water, many steamboats tied up at night rather than risk the hazard of the river. The Mississippi was master still when it would, but on the whole it dealt kindly with these new things, these curious impudent affairs of hissing steam and splashing. From the first they were of the great waters as different from the round-bellied, swaggering sea craft as could be. And in the dark night they paid sober worship, homage enough. The old Father of Waters was still a god to be treated with respect, a nightmare of a very convincing sort with which to spend the lean, black hours behind the spinning spokes of a steering wheel.

All up and down the Mississippi, steamboats were poking about in the dark, in the lower river, the low pressure, deeper draft boats of the eastern men, Fulton, Livingston and Roosevelt; on the upper the high pressure, flat-bottom boats begun by Captain Henry M. Shreve. The easterners had a fourteen year monopoly that made every steamboat builder pay them royalties on their invention, but Shreve did not let them keep it. He fought them in the courts, beat them, and the Mississippi was free of monopoly for all time. Two hundred thirty were in use by 1836. It was he who changed the whole style of steamboats. The easterners used a deep draft boat, put their engines down in the hold and used low steam pressure. Their boats were not entirely successful. One lay upon her side on a mud bank for five months. Her draft was too great for the changeable water way. Her builders were not of the river, properly speaking. The reason for their building deep hulls was to house the engines adequately and to save space. Shreve was a westerner. He knew the Mississippi and he had ideas. Years before he had been a flatboatman. He had learned steamboating on the old *Enterprise*. He had known the river at all stages. For him there was only one hull fit to navigate the great stream, a hull that would draw a few inches, a broad, flatboat hull sharpened at either end. The matter of the engines he handled simply. Since there was no room to hide them in the shoal hull he put them on the first main deck and stowed his cargo in the hold beneath. His boilers were built to make steam at high pressure by the improvement of the draft of his funnels. To test his theory he built the *Washington,* built her one hundred forty-eight feet long and two decks high. He put into her an engine with a cylinder twenty-four inches in diameter. His piston had a six foot stroke and she beat every boat she met. Seamen at New Orleans frowned upon her as unseaworthy. She was awkward to their eyes and a freak craft, but she made a round trip from Louisville to the Gulf in forty-one days, climbing the Falls of the Ohio easily. She was a pioneer and a brilliant one, for within a few years her type became the usual one for river boats. She did not live long. Shreve had not designed his boilers strong enough to stand the increased pressure. On June 9, 1817, under a full head of steam, there was an explosion. The superstruction was badly torn, the vessel was wrapped for a moment in a white cloud, scalding and suffocating, and since the boiler had burst in all directions the sheared beams of the hull gave up, the skin opening like an egg shell to the water which swallowed her. She was a total loss. Low pressure owners used her as a horrible example, and for many years those who sought safety travelled by low pressure boats. The start had been made, however, and the high pressure boat offered the argument of speed. Western life then was filled with risks. A boiler explosion was only an added menace, and the traveller in general gave the matter little consideration.

The best thought of rivermen went into the steamboat. Every hull launched was an added competitor. Boats have always been things of both love and profit. From plain hulls of simplest accommodations they became craft of great pretense. Captains came to have reputations, and people travelled by choice with a man who won through in spite of low water, lightning, drift ice, gales of wind and failure of his engines. There was no place for figureheads on the rather clumsy hulls. Many landings were made head on and so the forward part of the deck was kept clear for handling freight. Marks of distinction were, however, desirable. The funnels usually went straight up from their sides, reared high above the boat to give good draft to the fires, supported by struts from funnel to funnel, guyed staunchly fore and aft. From the side there seemed to be but one tall pipe, often banded with red or white. From ahead the two stood abreast each other like two strange towers belching black smoke. Between them was the logical place to display the marks of recognition to serve in place of the absent figurehead. The *Natchez,* swung upon the funnel struts a bale of cotton. Others carried crosses, circles, anchors, spheres, anything which pleased the eye of the captain and served to distinguish his boat at a distance from all the others. In addition, some displayed emblems upon their paddle boxes and pilot houses. These were painted in gay colors to catch the eye and were limited only by the fancy of the captain. In a broad sense they prepared the way for the gilt and tinsel ornament and the scroll-saw work which they sported later. Some became as bad as a wedding-cake or a Christmas-tree palace, tricked out in gauds of all sorts designed to give the impression of luxury and opulence. In time, carpets to the value of five thousand dollars were put on board and chandeliers of cut glass pendants and prisms were installed. Beds were substituted for bunks and cabins became spacious. Men such as Billy King developed hull lines that were faster and faster. He designed the *J. M. White,* the fastest boat that ever travelled the river. He found a boat made two distinct waves; one where the bow thrust aside the water, like the turning back of a plough-furrow and another, well down the boat's side, where the hull was trying to leave the water. Both of these waves have been carefully studied by the towing of scale models at a known rate of speed in the test tanks of our day, but Billy King merely saw that second wave and thought of the grip it would give his paddle blades upon the water if he moved his wheels aft to dip into it. To think was to do, and Billy King designed the *J. M. White* with her wheels set to use that wave. She left New Orleans and reached St. Louis, up-stream and against the current all the way, in just under four days. Four days to cover what for La Salle and Marquette was the effort of a lifetime, that was the saga of the steamboat, four days of comfort, flying at a pace that was miraculous! That quieted the mockery of salt water sailors, and the scoffer who decried high pressure steam. No craft ever matched that record. It stands as a monument to Billy King.

What was a Mississippi steamboat? To look at a picture of one is to get a glimpse of speed, of huge deck cargoes and clumsy looking hulls; of smoke and a slender wisp of steam against the sky and little more. The steamboats were part and parcel of the river they travelled. They were never standardized, never built over one mould. Their life was too short within that of the nation to show just what they might have become. They died of the railroads about the time the proud clipper ship disappeared from the sea. There were some things that all steamboats built after 1830 had, certain arrangements that were more or less uniform. Any river man could tell the boats apart as far as he could see them. He needed no recognition marks, no device suspended from funnel guys, nor emblazoned on paddle boxes. To his eye there were differences that were readily discernible. He had but to look to pronounce the far-off boat the *Enterprise,* or the *Comet,* or the *Paragon* or the ill fated *Sultana.* Yet they all had a tribal likeness, they all belonged to the great family of the Mississippi that grew out of the modest *New Orleans* and that reached to the glory of the *Robert E. Lee* and the *Eclipse.* By the days of the *Lee,* 1870, steamboats were three hundred feet long and forty-four wide. Engine cylinders were forty inches in diameter and bored practically true, as compared with the early cylinders of twenty-four inches truly ground only within a quarter of an inch. The *Lee* had eight boilers, her paddle wheels were thirty-eight feet in diameter and their buckets were sixteen and a half feet long. The infant, blunt-bowed craft had grown tremendously and were easy and sleek of line. Some cost as much as two hundred thousand dollars and most lived no longer than five years. Five years' work and bustle, battling the hazards of flood and low water was a hard life, but they were usually years of profit, boldly undertaken and handsomely paid.

What were the steamboats that used to be herded by hundreds along the levee at New Orleans, or Natchez, or St. Louis? Their hulls were shoal, sharpened at either end, built sturdily of tried timbers and planked to resist the thrust of snags and the strain of stranding upon sand bars. The holds formed by the hulls were filled with cargo, usually freight to be carried for the whole trip where possible, for it was difficult to handle the lading and get it smartly overside from the hold. On top of this, well supported, came the first or main deck. There, often upon brick foundations, were set the boilers, the engines and the kitchens. Travellers known as deck passengers were carried upon the main. The accommodations were crude, and all available space was crowded with freight, short-term stuff, to be put ashore at early landings and replaced by other going further. It was the realm of the deck hands and the roustabouts who hustled the cargo on and off the boat. Seldom clean, often noisy, travel there was rather disturbing but economical. Still, a deck passenger had to be something of a hardy creature, patient and long suffering, and not over nice. He lived amid a clutter of gear, lines, snubbing posts and unnamable contraptions.

Above the main deck rose the boiler deck. This could be gained from the main deck by a staircase forward. It was here the better class of voyagers lived. They came on board over the long gangways lowered by tackles to the bank, and usually went up at once to their own quarters, out of the undesirable main deck. The cabins here were good, the ship's office at hand. Frequently the bar room was at the forward end of the deck, a block of staterooms next, and in the stern the women's cabins. There was a footway entirely around this deck, in some boats, meriting the name of promenade, in others the merest passage between the house and the rail.

The roof of this deck provided what was known as

THE STEAMBOATS WERE HERDED BY HUNDREDS

the Texas or hurricane deck. It was accessible to passengers, but was given over to cabins for boat officers only. Save for an isolated block of these cabins, it was a wide expanse of foot room. Modern sea-going steamers are often rated as desirable or undesirable by the number of funnels they carry. The more funnels the more fame seems to be the rule. This is so pronounced in bringing a vessel into favor that even a giant such as our own *Leviathan* carries a dummy, used to conceal tanks, in order to give her the reputation of having three funnels. The Mississippi steamboat advertised her Texas in the same way. The hurricane deck was a point of honor, a badge of merit which insured to the passenger he was travelling on a steamer that warranted his pride in her.

On top of the officers' cabins, above all the boat, lifted the pilot house in lonely splendor. This was often an elaborate affair tricked out in gingerbread ornament, gaudy in resplendent paint. It looked much like a summer pavilion from a great estate, stuck atop the high-sided boat. Aft of it, one on either side, were the exhaust pipes which, when under way, gave off a steady curl of steam, white against the sky. Forward of the house the two funnels towered above all, their tops frequently ornamented by serrations till each seemed to bear a king's crown of black iron. From their throats poured smoke and sparks and at night, when the boat was travelling at full speed, a lurid, red cloud of fire and smut.

For the rest, each boat was a law unto itself. One

would carry its sounding boat on davits at the stern. Another would hoist it aloft to the hurricane after sea-going practice. Captains determined these things as they decided almost all else about the steamboats. Frequently a master owned his boat outright, or at least had a share in her. One captain decided to attract trade by a brass band. Another preferred the orchestra. To match this, that fearful and wondrous instrument the "steam piano" as placards announced the contrivance, was put on board. Many took passage to hear this instrument, but they found it a bad travelling companion, for its hooting was as easily done at three o'clock in the morning as at any other time, and there was no sleeping in the same boat with its din. One trip sufficed for a suffering passenger. He usually returned by another boat. Captains even vied with each other in the forms of their tickets and posters, trusting to elaborate printing, lettering decorated by swirling scrolls, or bold with heavy-faced type. The business was expensive. It could cost as high as eleven thousand five hundred dollars a month to operate a boat, and even more. Pilots sometimes cost five hundred, and a captain two hundred dollars. Wood for fuel at twenty-five cords a day could run up to two thousand dollars and food to as much more. No wonder every captain was on his toes to make his boat the best on the river. The competition was savage, but the profits were satisfactory at times. Net profits with all expenses paid totalled sometimes twelve thousand dollars a month, and on the upper river, between the going out of the ice and the return of winter, a certain service of five months, net earnings for a single boat could run to fifty-six thousand dollars.

Some boats carried the paddle wheels at the stern, but most were side-wheelers. The stern-wheelers had a larger main deck and therefore more space for deck loads. Two engines drove the wheels, both turning one shaft. The hulls were limber, and every ton of freight carried changed the alignment of the engines, but the stern-wheelers threshed along right merrily. They had one fault. In a tight place they could not be steered so easily, even with their compound rudders working in the currents set up by the churning wheel. They were more picturesque to look at, but not nearly so easily handled. It was impossible to turn them sharply, for the wheel could not be made to aid the rudders.

The side-wheelers used two engines, each turning one wheel. Far below the pilot house, in the gloom of the main deck the chief handled the starboard or right hand engine, his cub stood at the throttle of the one on the left. Amid the eternal smell of scorching oil and escaping steam these two obeyed the pilot's bells, working their engines by hand, when called on to do so. By going ahead on one engine and reversing the other, a side wheeler could be turned in her own length with practically no advance. But underway, even a side-wheeler was hard to stop. After the engines were halted the sliding boat was dragging at least one bucket submerged to its full depth of three feet, and two others partly submerged. To turn these wheels against the thrust of the hundred or more tons of boat sliding through the water at nine miles was an impossibility. The boat had first to slow down. To turn the wheels at the earliest possible moment was to give the pilot power to handle the boat more quickly and to guide her more surely. For this the engines used a "club," a wooden stick, two or two and a half inches square. This was jammed between the rocker arm and the lever, which lifted the inlet valve, admitting swiftly, while it was in place, an enormously increased pressure to the piston, permitting the engine to turn the wheels despite the thrust of the water upon them. It was dangerous but it was necessary. Sometimes a cylinder head would shear off its bolts and blow out, scalding the engineers. Sometimes the paddle wheel would break under the strain, and the pieces, wedged against the paddle boxes, which covered the wheels, would bash and splinter them, crippling the boat. Usually, nothing happened, and the pilot gained the power to maneuver skilfully. In any event the club was in constant use, the engineers taking the risk as they took so many others of their calling. Speed was the point, speed and power, and it was their business to supply both. When anything broke the engineer was enough of a blacksmith to replace any broken part. It was all in the day's work. Anything was forgiven but catching an engine on its dead centre, where it could turn neither forward nor backward. This was the terror of the cub perspiring at his throttle, greasy, dirty, with that awful possibility staring him in the face. It meant hard work to get an engine off its dead centre and the efforts of three or more men. There was enough work in that life of steam and grease without carelessness adding to the burden. A cub who centered an engine was unpopular, and to be unpopular among the rough crew of a steamboat was a wretched state. He was hazed unmercifully.

The pilots were responsible for guiding the boat but for speed they looked to the engine room. Hot steam could be made easier from a little water than from much. If the water got too low in the boilers the steam was made so fast that the boiler exploded, frequently killing all on board, blowing off the hurricane deck, perhaps shattering the bottom. For speed an engineer carried as little water in his boilers as he dared and engineers were daring men. Life was cheap on the Mississippi. There were so many ways a man could die in the new country that the hazards of a steamboat were taken lightly, and reckless rather than sane practice was the rule.

Many engineers not only made steam with the water as low as possible, but they tied the safety valve shut so that it could not blow off and relieve the pressure. Thus bottled, the boiler was compelled to stand stresses for which it was never designed. If luck were with the engineer his engines ran better than they knew how, but if luck were bad the rivets surrendered, the tubes gave way, and in the sudden expansion of the steam the unfortunate boat was rent from end to end. Men were blown through solid wooden bulkheads, were lifted from the pilot houses and thrown a hundred feet, and lived unhurt, like the survivors of a freak hurricane, but most of those on board a steamboat whose boiler burst died at once, or were horribly scalded or crushed. Yet the engine room furnished the speed the pilots needed and crew and passengers gambled their lives upon the cunning of these little-seen men, who lived below in the stench of oil and steam.

When the hazards of ordinary struggle with the river are considered, with their risks and dangers, it is remarkable that steamboatmen paid any heed to racing. Men of no more than average acumen and energy would have found themselves busy enough without the added strain of trying to pass a rival. Both the upper and lower river saw this sport of madness, this striving for speed with life and limb in the balance. As a matter of fact there were hard and solid reasons behind the reckless efforts. The vessel which sailed close to another schedule won the trade only by calling first and carrying off the freight from under the competitor's nose. Only where such warrants for the practice were evident would owners listen to races, and then only if the trade in question were worth it. Yet there was something in the trial for speed that caught the blood. A passing steamboat lashed even timid passengers into a fury. No one liked to be so left behind. A shipper of barrel staves would give the captain permission to burn his share of the cargo, or an old lady from Kentucky would dedicate her tubs of lard to the making of sudden steam to let the boat on which she rode nose past another, both tearing madly along the river. There was much noise and vainglorious strutting over southern races, when the levees at New Orleans were crowded with steamboats. They usually sailed at five o'clock of an evening for up-river points, all gilt and color under their smoke clouds, ploughing up-stream bravely. In the north, racing was casual but still very interesting. Fewer roustabouts hung about making reputations by their talk, but the north had a lore of its own, a similar tale of flying hulls and hissing steam.

There was a great boat in the north. Her funnels were bell topped and striped with paint in a coronet pattern. Her hurricane deck was elaborately railed and the lattice of her railing on the boiler deck was no mere picket fence or cattle rail. No sir, it was artistic that rail, all points and spikes, and the palings of her paddle boxes also. *Grey Eagle* was her name and she bore it boldly painted upon all four sides of her pilot house. On the paddle boxes a great bird flew across a brightly painted landscape. Her bow had a beautiful flare and she was trim from bow to stern, her white paint clean and well kept. She was almost new in 1856, just old enough for her captain to know her, for her engines to work themselves into easy running, and her pilots to outguess her. Captain D. Smith Harris was proud of her. He had built her with his own money, and she had cost sixty thousand dollars. The pride of the river he had wanted and the pride of the river he had. She was queen of the upper stretches.

In 1856 the great under sea cable had just been laid. England and the United States were connected at the tap of a key. The broad Atlantic was wiped out by the ingenuity of man. To celebrate the event Queen Victoria sent a message of commemoration to President Buchanan, proving both the success of the cable and the good feeling between the nations. There was no telegraph line to St. Paul, but there was to Dunleith where Captain Harris was loading the *Grey Eagle* for the run up-river. Sixty-one miles to the north, and therefore sixty-one miles nearer St. Paul, another telegraph line ended at Prairie du Chien. There lay the *Itasca,* also loading for St. Paul under Captain Whitten. Both boats got the news of Queen Victoria's message. Captain Harris decided to be the first to reach St. Paul with it. A whim, a vagary, if you will, but truly a touch of old romance, of a desire to partake in the great event of the decade.

Harris slipped out quietly. He knew what he was about and his preparations were sound. First, he carried enough good, soft coal, the best the market offered, for his fires. As a reserve, against the hour of need, he had ready some barrels of pitch. On leaving Dunleith the captain saw to the storing of his holds so as to trim the *Grey Eagle* for her best running. No more cargo would be taken. Whatever he carried billed for river landings on the way up he would drop on the return, taking it on first to St. Paul. Did he not bear the Queen's message? He would brook no delay. One obstacle stood in his way of making the run without hesitating. The *Grey Eagle* carried the mail. Since she would be so far ahead of her usual schedule there could be neither freight nor mail to pick up, but delivering the mail she carried was a problem. Captain Harris rigged out one of the long gangways the ship carried, so that when she ran close to the bank a man, perched on top of the tilted gangway, could drop the bags safely to the bank below him, without the boat coming to a full stop. Thus prepared for a fast run the *Grey Eagle* slipped out into the stream in an effort to cut down a sixty-one mile lead.

Captain Whitten had in the *Itasca* a twelve-knot boat. At his usual rate he had only to run two hundred twenty-one miles while the *Grey Eagle* was doing two hundred ninety. This meant, that to win, the intrepid Captain Harris would have to average more than sixteen knots, a tremendous speed to maintain for a day and half a night. Eighteen hours of flying would win through. Captain Harris spared neither boat nor men. It was a day when firemen were encouraged to work by a generous supply of whiskey put before them. Men who drank were supposed to shovel with renewed vim, catching the spirit of the thing as it were. Up the river roared the *Grey Eagle,* everything wide open. The furnaces were hot through the jackets. The funnels belched flame as well as smoke. Men were stationed on the hurricane deck with hoses to wet down the funnel breechings and keep the decks from catching fire.

Captain Harris was a driver, a driver gone mad with an idea. He would catch the *Itasca* whether she would or no. Miles flew past. The boat surprised even her captain. The engines turned sweetly. Sixteen knots, sixteen knots, sixteen knots and more, she sang. Over

227

and over drove the paddles. The deck hands were astonished. What had gotten into the old man? Where did he mean to fetch up? Along the banks he shot, dropping mail sacks, waving to a stray hanger-on along the dock side. The bell jingled for full steam again. The boat veered off into deeper water. Sixteen knots, sixteen knots, the engines sang. And still no sign of the *Itasca.*

Soon there was no freight to be seen on the piers or along the levee. The *Itasca* was cleaning up and could not be far ahead. The sixty-one miles were being cut down, but St. Paul was drawing near. The Mississippi is crooked on those upper waters. Islands melt into other islands. Bends sharpen and long sand spits make off between the bluffs. To be hung up there would end the race. The *Grey Eagle* pilots kept to good water. The old man would take no excuses on a run like that. Miles vanished, succeeded by other miles, in a long, never-ending sequence. At last they rounded into an open stretch. They were close to St. Paul now. Ahead in plain view was the *Itasca.*

Captain Whitten saw the *Grey Eagle* as soon as her people spied the *Itasca.* He knew in a flash what was up. The *Grey Eagle* was so far ahead of her schedule there was no mistaking the matter. A race! A race then, and St. Paul just ahead. The *Itasca* tied down its safety valve and leaped ahead. It was time for Captain Harris to use his pitch. The fires roared up the flues. Great black clouds hovered above her funnel tops. Both steamers were under a full head of steam as attested by

THE *ITASCA* TIED DOWN ITS SAFETY VALVE AND LEAPED AHEAD

the white wisps that whined from their exhaust lines, high above the Texas of each craft.

On and on they drove. Harris gained slightly, but the city was just ahead. Would the *Grey Eagle* stand it? Dare the engineers try her boilers more? She had already suffered much on the long run. Could she endure? Captain Harris nodded yes. They crept up toward the *Itasca*. A stern chase is a long chase, but the gap was closing. Now they were in her wake riled to a frothing white of torn water. Now they were on her stern. Once up with her the *Grey Eagle* forged rapidly ahead, but the *Itasca* responded grimly. It would be a fight to the levee, a fight to the taking of the lines.

Captain Harris was on deck now. He wrapped a bit of paper, on which was written the message of Queen Victoria, about a chunk of coal. Slowly his boat won its way. Foot by foot she crept along the sides of her rival. At last she was abreast but the levee was close. There was a crowd there waiting for the *Itasca* which was very close to her schedule. A roar, a race! Hats went up. Puppet figures, urged by excitement danced on the bank. The boats were neck and neck, the two prows sending up their fountains side by side. Then Harris got his boat out into the lead, foaming along the levee. The wake waves splashed and washed at the bank. A foot, a yard, a quarter length she gained. Harris gauged his distance and tossed his weighted message upon the levee where it was snatched and read. He had won, won by a quarter length. Cheers went up.

231

Excitement was high. Whistles blew, bells clanged. Harris was happy. The *Grey Eagle* had made her average of sixteen knots. She was entitled to hoist the broom, emblematic of having swept the river. A man climbed up and fixed it to the finial atop the pilot house. Whitten was beaten. Of the score of boats Harris had owned the *Grey Eagle* was the sweetest. Up-stream running held no terror for her. She had averaged actually sixteen and one-ninth miles for every hour of her run of two hundred ninety miles. The *Itasca* had averaged twelve and two-thirds. The handicap of sixty-one miles had availed nothing. Another chapter had been added to the history of the Father of Waters. All hail the Mississippi! All hail to the *Grey Eagle!*

There were many boats which possessed the mystery of speed. The builders could never tell surely. It was a God-given quality. The *Key City* and the *Itasca* were built to the same lines. Their boilers were of equal capacity and similar design. Their engines were duplicates even to the point of their parts being interchangeable, which in that day was unusual. Yet when each was at her best the *Key City* could leave the *Itasca* three miles behind in an hour's run. She had the magic touch, the rare perfection, perhaps put there by willing hands at her building, or possibly an accident of a twisted frame or a mould that slipped. Speed was not even limited to side-wheelers. The *Messenger* on the lower river was a flyer in her day yet she was a stern-wheel boat. Every craft was an experiment, a groping for an ideal, a subtle truth to be understood and tried in the face of the old Mississippi. No wonder Captain Harris was happy in the possession of the *Grey Eagle*. Even the men who had worked on her could not tell what they had done to breathe into her the secret of her wings. She was a jewel, a miracle. River men were rough and careless, but they could appreciate such a rarity. She was a great boat, the *Grey Eagle,* six hundred seventy-three tons of marvel, the queen of the upper river.

CHAPTER ELEVEN

RIVER GODS

A SOMBRE river, the Mississippi, save when the
sun gilded its grey water and made its far-flung
banks into a fairy land of green and browns and yel-
lows. A river that called for gods, that clamored for
deities to be worshipped. It had them, strange water-
beings, malevolent, heedless of all but themselves. They
were creations of evil for most part, sinister and violent.
Only gods able to live in the wilderness could fit their
destinies to the river. Of need they were free and strong,
savage as the floods, enduring as the mid-ocean current
that did not change, variable as the eddies that filled the
bayous and chutes with horror and dread. Strange gods
of barbarism whose hearts were steel and were strong
beyond the strength of men. Gods who had something
of the earthquake and the hurricane in their fibre, unknown in motive as the blast of tortured volcanoes,
given to tornado freakishness, fancies of the moment,
shifting purposes. River Gods!

Like shadows they moved among the first white men
to ply the river and settle along its banks. They preyed
upon pioneer efforts, robbing ruthlessly, slaying fantas-
tically, inspiring the same awe as the ungovernable
stream. They were things of evil, which the immigrant
could never appease. Today they are but legends,

234

legends that terrify even in memory. Beginning as brutes and men in a curious motley, they became puffed and swelled until their might filled the sky, weighing upon the hearts of river folks with a ghoulish, unnatural fear, a horror like that inspired by beasts of prey so powerful men trembled at mention of their names. These gods were often short lived, gods of violence, who were frequently swallowed suddenly by the silence of the river from which they had sprung. It was their mother, so much stronger, so eternal, that they were but shadows across its face, fleeting shadows, but grim and awful. They were spirits fitted to the mystery of the great vastness, water ghosts which lived convincingly enough in the half light of the early days.

These gods were greater than history, they were legend. They sprang from the slime of the river, brooding in the bluff caves like alligators, but they travelled like horses, striking their prey swiftly and sweeping on in their careers of madness. They played the part of fate to many, cruel and cunning, so primitive they never knew remorse, so dastardly in their acts that the relation of them chills the blood. Some lived long enough to find greater opportunities when the steamboats came. Most perished before, going back to the river that bred them. Such an evil-omened god was Mike Fink.

Mike Fink is a legend and he still lives in the minds of the old rivermen. He was godlike in only one thing; he was strong. His bullet head was hard and his prodigious lower jaw jutted determinedly. He may have come forth in good earnest from the river bed, for no one knows his parentage. A man so hardy would have little need for father and mother. No rearing could have touched him. From the first he was destined to be a river god. He began, modestly enough as a scout. For a time he travelled the woods, escorting settlers to their new homes, protecting them from Indian attacks, supplying for their food, game which he killed adroitly. His rifle was good and his eye sure. Scouting, however, gave him little reputation and much work. Moreover he rubbed elbows with duty at every turn. There was no glory in shooting Indians who strayed across his path. Beside, there were orders and restrictions which to Mike meant monotony, so he quit and went afloat upon the Mississippi as a keel-boatman. The rifle went with him and he kept his eye good by shooting an occasional negro he saw along the bank. "The Snag" they called him and there was many a man who run afoul him who found the name's full meaning. After all he was childish, but as a fighter he was without peer. "I'm the son of lightning," he would shout, "I'm sudden death and destruction. There's no cave can hide you, nor rock that can kill me. When I come up to you you can't move, wow!" Then to prove his cry he would begin to take apart any man upon whom his gleaming eye rested. There was never such a man for mauling. His hands were terrible. He could tear loose a man's floating ribs from the breast bone with his fingers, unaided. There was no black or white on the river whose skull he could not crush by butting with his head. His knees were as good as his hands in a rough and tumble fight. No

man's ears were safe from his teeth, and the blow of
his fist was like the kick of a mule, while a prod from
his heavy boots was sure death. Among men who were
as rough as any in the world he stood head and shoulders above all, filled with the crude strength of the
river; yet he came to an early end.

Mike Fink was sociable, and to be sociable meant
that one spent one's evenings in drinking houses reek-
ing with the stench of crowding men, the smoke so
heavy as to seem solid, the talk raucous and loud, voice
clamoring with voice to be heard. Mike ruled such

evenings and was a regal king. His strength made it safe amusement. The attack of a mere half dozen fighting rivermen could not frighten him. Others he saw stretched upon the floor, but for him there was no dread; yet he died young.

The god had a weakness. He fell in love, and his love was but half returned. Sharing of anything was against the grain of the champion, so he grew savage, using men with undue roughness, hunting trouble for the thrill of it, doubling his fame. The lady in question remained coy in spite of his behavior. She was not to be won by such simplicity. Instead she fawned upon a rival boatman. Through the long days on board his keel-boat, plying the river, Mike brooded upon his plight. To banish it, at night, he drank harder than ever, and cracked a few more heads for diversion. Then, one night, after he had warmed to the bite of the raw liquor he caught up with his trouble in one of the thousand drinking houses where he lived.

His rival, upon whom the lady had showered her obvious attentions, came in to pass the evening. Rivermen took their whiskey straight, four fingers deep, and at a gulp. There had been several rounds and all were becoming somewhat heated when Mike Fink developed a fiendish scheme for putting his troubles at peace. Drinking seemed banal. It needed the exercise of a little art, a heightening of the charm. A novelty would liven the evening. The rivermen were set for the shooting of the tin cup. In this sport a man took a tin cup, filled it with whiskey, a bumper drink, and stepping over to the wall set it on his head. A friend would then take up his stand at the other end of the room with a rifle and shoot a hole in the tin cup. The liquor, thus tapped, would flow out in a fountain to be licked by the tongue of the man upon whose head the cup was balanced. Shouts rose, merriment reigned at the fellow's grimaces as he caught and swallowed all he could of his liquor bath. It ran down his face, in at the throat of his shirt and across his hairy chest. Then everyone lined up at the bar to celebrate the sport until another cup was to be shot.

A primitive cunning crept into the eyes of Mike Fink. His was not a prepossessing face. The eyes were set too close together, and the flush of the whiskey mantled his face to an unholy red. He was less touched by the drink than the rest; that he owed to his strength. Loudly he called for a tin cup to be filled. His rival was hustled up against the wall and the cup set atop his head. Mike took his rifle and went to the other wall. He aimed carefully and squeezed his trigger slowly. The rival fell at the report, twitched once or twice, and lay still. The tin cup dropped to the floor and rolled away, untouched by the bullet. Mike had drilled his rival neatly through the head. He laughed in the silence that had fallen upon the men, a roar of a laugh. The lady would trifle with him? He cried to the drawers to set up drinks for the house, but that round was never begun.

The slain rival had a brother among the crowd, who knew Mike and his rifle. That man glanced once at the dead man and again at the laughing Mike, proud in his strength, gloating over his fancied cleverness and perhaps a little surprised at the ease with which the thing had been done. There he stood, a demon of glee. The brother knew better than to chance hand grips. He threw up his own rifle swiftly, and took Mike where he stood overtaken by river justice. At the shot, the laughing face stiffened, looked more surprised, and the great body crashed to the floor beside that of the rival. Mike Fink was neatly shot through the head. The favor of the lady interested him no more. There was no more terror of the great hands, the stony head, the ramming knees, the deadly feet. The river had lost a god.

Many gods were born to the great river. All down its length men grew to be well known for their ability to be rougher than rough, but none, for strength alone, exceeded the repute of Mike Fink. Historians were lacking in the early days. A man's tongue wagged of something he had seen. Another heard, enlarged upon the relation and undertook his own account. It was by such accidental news that the great were made, that time was prevented from swallowing them and that Mike Fink has lived. Not all the accounts were of gods so simple minded.

Legend created wonders out of nothing. Darkness and fear bred superstition and the early river people had more than enough of both. Report gave curious power to some wreckers. There were tales of a one-eared inn keeper who was closer in touch with the devil than was fashionable in his day, and who therefore had a very sinister reputation. Keel-boatmen could be met about the river, of a night, who were no more responsible to the laws of nature than the demon inn keeper, or the ghostly wreckers. These ghosts sailed usually in the teeth of the wind and current and vanished in a clap of thunder. All the many caves that river and weather had fashioned in the bluffs had their own bogies and their own tales fit to turn the blood. The whole valley was given over to the worship of these terrors as being keyed to the daylight reverence for the gods. Not all the deities they gazed upon with wonder were so primitive as Mike Fink. More subtle beings moved abroad, who used their minds, and whose wits were sharper than most people's had any right to be. Of such was John A. Murrell the giant of the lost legions who lorded it along the river. The settlers were for most part people of staid habits willing to work, anxious to acquire wealth. Their sense of locality was great. Given success, there was no more moving, they remained where they were until their death. Among them flitted the river gods, brave enough and clever, but often utterly without mercy. Murrell was true to his roguing kind. What was worse for his victims, he was not a vulgar lout making a bar room ring with his roving jargon, but a man able to inspire respect. He was easily the most dangerous god the river bred.

As little more than a boy he came out of Tennessee and began in a modest way. Catching a boat unguarded he carried off what he could. Emboldened by success he undertook highway robbery. Boatmen were still paid off at New Orleans and left to their own devices to find

their way home. Murrell was a fine-looking man, tall and powerful. His personality made him a good travelling companion until it was too late for the boatman to save himself. In every such trek upon the river there came a time when the boatman would find himself looking into the barrel of a fine, rifled pistol backed by the coldest eyes in the world, the eyes of Murrell the killer. There might be a moment in which the trapped wretch could ask for time to say his prayers, or begin his plea for mercy, but no more, for Murrell was a man of methods. Whom he robbed he killed, dispassionately, as the faun or satyr might have slain a human strolling in the glades of old Greece, when Zeus ruled on Olympus.

There was a feeble legal justice on the river beginning to pay attention to violent crimes. Local courts judged such matters as came before them, but they were limited by their inability to carry out their sentences. Their findings had to be based on at least some shreds of evidence and this was the very thing that young Murrell never gave them. He shot his man first, on the whole politely, then he robbed him at leisure and when he was finished he disposed of the body so thoroughly that no victim ever embarrassed him. The river took care of its son. It received the body, after Murrell had disembowelled his victim. This horrible business prevented the body from floating as decomposition took place. With the body were deposited the murdered man's papers and effects. If the unfortunate had any clothing that pleased Murrell better than his own this was put on at once, and the murderer carefully gave his own discarded garment to the river. There were no clews left to lead to him; Murrell was brainy enough for that.

Rivermen however, had a justice which took but little heed of formal evidence. That justice had no prosecution nor defense, no tomes of wisdom, nor endless bits of precedents. Instead, it began in the low rumble of angry voices, the lurid flash of torch light, high shrieks of half-crazed fear, the crackling whirr of rope dragged on a run over the dry bark of a tree limb, the soughing creak as a body slumped deadly upon the

243

hempen fibre, a fusilade of shots fired into something that swung writhing under the groined forest roof, the whisper of departure as lynch-law ended its course. For such as Murrell these practices presented the greatest danger, but the long man from Tennessee was better than any mob. He understood them. Standing afar off he saw men coming to his river in ever-increasing numbers. He put his finger upon the pulse of this host of mobs and learned that the heart beat quickened most to fear. Weary of single murders at the point of his rifled pistol he took time to think this over. He knew steamboats were coming more and more to western waters. Things were not so isolated. His chances of detection were greater. Should he fail, lynch-law was at hand. Thus revolving the current state of affairs within his mind he came to two conclusions. He would give up the single-handed banditry out of respect for lynch-law, and he would operate in spite of this new mass justice. He would lead men into new undertakings. These steamboats had made his old game unsafe, good; it offered new opportunities for more interesting villainy. If he could find a fear greater than that for lynch-law, he would be able to hurl every rogue on the river into action. There was no limit to the profits. He would be Murrell the dictator, with the wealth of the rising river trade flowing into his coffers. So he took a trip to New Orleans to think it over.

He saw much, and everything he saw he weighed to find his idea in it, the thing which would raise him from a single bandit to the master of the Mississippi.

Among other things he visited camp meetings, where men and women gathered from miles around. He looked on their worship with a cold eye, weighing their enthusiasm. They were uplifted by a common centre of religious feeling. They belonged. They all had similar experience, save a few who stood exalted. Those few were at the heart of the thing, giving it its fire as they felt it. All were afraid of the displeasure of God, and of each other. To Murrell's pagan eyes there was nothing else there, no humble honesty seeking to get closer to heaven, no simple mind pleading to understand the ways of God, no devotion carried away into ecstasy. These things the river god regarded not at all, looking with his hard eyes upon these camp meeting folk. He saw, however, the power of enthusiasm coupled with the fear of eternal death, and out of that experience, colored with so much of good, he forged the unholy scheme of the Mystic Clan.

He used a religious approach and became a preacher. The leaders of those early camp meetings were vigorous, hard-working men. What they lacked in learning and wisdom they made up by enthusiasm. Under the blasts of their fiery zest listeners were melted to tears. Sorrow for sin shook their audience, regret, remorse, a desire for godliness stalked the aisles. God was made very real for those who came seeking Him in the wilderness, turning aside from their daily work. There was much soul searching and many heart throbs. The music was simple, like that of the negro spirituals. The voices were untutored, the addresses rough and overly emphatic. There was much to be desired, but for the most part a camp meeting was sincere. It meant well. With tears streaming down their faces men and women decided to do better things, to live cleaner lives.

John A. Murrell faced those men. He wrung their hearts. His language was fitted to the part. He exhorted. Men stared upon him, drinking in his words. They swayed to the play of his thoughts. Pulses quickened, faces lightened. They were his, his to turn and direct. God lived in such a man they thought, and through it all not one body rose from the river mud to scream "murder." Not one doubt faced or checked his progress. They took him into their homes, this powerful, faunlike man who respected neither life nor death, who in his heart was untouched by any desire but the building of his Mystic Clan. There he chose whom he wanted for his followers, finding time between the meetings to pick and win those who would serve his purpose.

Upon the crest of religious fervor he tested and tried them. Cold and deadly of purpose he led the folk of the river town to form a curiously dark and sinister fellowship. A few, those he felt most surely to be of service, he admitted to an inner council, an exalted circle within the larger union. To those especially chosen passed the great work of organizing. Even when he met them as friends, even as he spoke from the platform with all the heat and soul-stirring vigor he could so well bring into play, Murrell never got very far from his fine, rifled pistol. Anything might happen. He was prepared for eventualities, but he would go no more upon the road as a lone bandit; he was playing for better stakes.

The Mystic Clan! What a name for Murrell's brotherhood of rascals, for it was only the immoral portion of his audiences that interested him, that was worthy to join in his scheme. There were attractive features. The clan was secret. It had pass words and grips. There was not too much of ceremony, for the river folk did not take kindly to such doings, but there was secrecy and there were dues. That was for the comfort of the brethren. Sometimes the clan hired lawyers to protect members from the law when clan business carried them to the wrong side of the jail gates.

Clan business was not difficult to understand. In a way they were in trade. They dealt in horses and slaves. The horse dealing was profitable in itself. A man cannot help but make money when he rides a stolen horse, which has cost him nothing, from one state into another where it is sold for a good price. The slave trade was less pleasant. A negro ran away from his master into the arms of the Clan. There he entered upon a contract to let them sell him to a new master. He shared in the money of the sale, and shortly ran away again, when the performance was repeated. The Clan thus sold negroes they had never owned and repeated the process at will. No one ever came closer to making money out of nothing than they. The sole investment they made was a slight risk. Even that they did not take honestly. When the negro fell into their hands after the fourth or fifth sale, and they dreaded inquiry into his actions, they murdered him in cold blood, destroyed his body, Murrell fashion. There was no witness left and no direction for suspicion to take. A prime lot of rascals and a dastardly scheme of operation!

Even within the clan there was injustice and double dealing. A thousand men belonged to Murrell and he divided them almost in half. One part he admitted to the inner council. Plans were made by them, the largest share of the profits was theirs and so long as they kept their ears open the risk was next to nothing. The second part, known as the "strikers" a compliment perhaps to their readiness to act, carried out the orders of the inner circle, committing theft, murder, breach of trust and a host of lesser crimes. In return they received but a small portion of the booty. Revolt was vain. Murrell had chosen his inner council with discretion. They were masters of the strikers who could be accused and tried at the expense of the courts of law, or quietly murdered at the will of those who planned.

In a thinly populated country the thousand men were well scattered and posted so as to help each other in the event of necessity. Again and again a member of the clan would be suspected, but his comrades, by careful work, could divert suspicion or if necessary carry the man off and hide him in the swamps and cane brakes. Murrell was in the saddle. He had beaten the steamboats. They had broken down the isolation, made communication between districts too rapid to be safe for the criminal, but they brought more and more people to the country, more horses to be stolen, more slaves to be sold and sold again. The Mystic Clan was a working organization. Murrell was everywhere, seen now here, now elsewhere, watching over the work of his subordinates. Knowing himself to be faithless he trusted no one.

Had he been content with his progress he might have continued a few years and left the valley an independent man, but that he could not do. He was no boatman, yet the river was in him. Its ceaseless current, the ever-changing banks, the rush of its omnipotent force, these bit into his very being. He could not restrain his restlessness. In the might and relentless energy of the Mississippi he saw the result of ages of struggle. His destiny was to simulate that eternal emotion, to plunge on and on. His favorite ground was in Arkansas where the river was no longer a rushing torrent, trivial, impetuous, but a moving sea. At flood times it could become forty miles wide, running over into the bottoms, lowlands that it possessed for a time. He would be as great.

His camp meeting tactics and the mysteries of his clan gave him power over the negroes. The slave minds were looking for a deliverer. He went abroad on horseback organizing a revolt of the slaves. Emperor of the blacks, the whole valley under his thumb, the city of New Orleans captured and ruled by his clan. Here was a dream he saw by flashes, jagged lightning strokes. At his back were a thousand men who knew him for lord. The negroes could be made into good fighters. While they captured towns, the higher officials of the Clan would be able to rob at their leisure. The wealth of the newly established banks would be their affair. Remorseless, savage, the liberated slaves might throw the valley into such chaos that the settlers would welcome the Clan as the only restorers of law and order, law and order at a price.

While the work went forward, Murrell thought it desirable to enlist aid for his undertaking. He knew its size and its dangers. All his life had been spent in such attempts, but never on the scale he was considering. For a time he vanished. He went to Mexico to arouse interest. There he met the Latin temperament. Vainly he attacked the hard-headed aloofness. Polite but hesitating, the Mexicans were very different from the simple, rugged river folk. The farther he got from the mouth of the river into the foreign territory, the less powerful his spell became. Camp meeting tactics got nowhere with a people who were skeptical of everything, even of each other. He met the Latin desire for delay. They would dally. He learned a new word, *Mañana,* the vague "tomorrow" of the Spanish Americans. They spoke of it politely with a shrug of their shoulders, a gleam of the eye, a wave of the hands, a formal bow. His heated words were chilled by it. He recoiled at last, rebuffed, and returned to his river where he was god. The Mexicans were not good fighters, he told his Clan, but what he really found he kept locked in his heart.

Even the gods of Olympus had their weaknesses. Achilles perished because of his vulnerable heel. The

250

river gods were not immune. Mike Fink met his end over a woman, and even John A. Murrell had his failing. It was an outgrowth of his love for words; fine, large words which he spoke roundly. At the camp meetings he had used his power of speech with great effect. As he recruited new members for his clan it became a habit with him. He opened chapters in his personal history which pleased him to relate. He accented the safety of the organization by revealing its murderous removal of any evidence that threatened it. Most minds did not see the veiled contradiction that lay in such procedure. Murder and safety had never before been expected to live so happily together. Murrell made their association seem natural. That was the power of the man.

It happened that he tried to win to his cause Virgil Stewart. Stewart had a clear head and Murrell worked hard to win him. He was apparently successful. Stewart took the prodigious oath of the Clan, and in a short time Murrell had him made one of the inner council that his keen mind might aid in the planning of the great revolt. He was consulted as to how the blacks could be organized as a unit to rise against their masters. How could New Orleans best be approached? How quickly could the Clan begin to govern the territory once it was captured? These and a thousand considerations like them rose for the council to decide, and Stewart helped answer them. Murrell, drawn to the young man, took him about to aid in the work, and on their travels the river god revealed his weakness. He boasted. With the young man as a sole audience he poured out his soul, glorifying his ego. His tales dealt with frogs croaking above the skeletons of his dead, of shooting men through the back of the head and stealing their boots, of killing men with a leaden whip butt, and startling the night with the shooting of a trusting negro out of caprice. Stewart listened carefully. Murrell's liking for the young man grew. He talked of his power and might. He gave names of confederates, names that startled, for their owners were powerful, reputable people in the valley, with no blemish of rascality upon them. They were scattered up and down the river. What was worse, Stewart's travelling with Murrell gave him a chance to confirm his leader's boasting. It was true.

When that fact was borne in upon Stewart he listened no more. He had caught a glimpse of the red horror within which he was entangled. Murrell seemed an angel of evil, the recital of his deeds resounded through the brain of young Stewart like some bestial litany intoned by a demon. The recruit's ears rang with it, a hateful, maddening pronouncement of a man at war with his kind. And what a war! A faithless, brutal account of inhuman actions. Stewart looked upon his leader as upon a bogie, a monster out of a nightmare. He had lived all his life along the river and he was not unused to rough life and heedless action, but the callous cruelty of Murrell plus his devilish ingenuity in selecting just the people who could best serve his purpose decided Stewart. The last of the river gods had made his mistake and his doom was upon him.

Virgil Stewart broke his oath. He published all he had learned of the Mystic Clan. He gave the names of those of the inner council. Murrell tried to have him murdered. Decent people would not believe him. His list of names touched those who stood high along the river, too high for his charges to reach them. Instead, Stewart was rebuked for breaking his oath. Had he no sense of the solemnity of his plighted word? What did the fellow mean by behaving so? He was a perjurer. Had he not broken his oath? How could his charges be honest?

Murrell was arrested and stood trial. His Clan took the arrest as a signal for the rebellion in his favor, but without the river god there was no mind to guide and the attempt was stopped summarily. All before had felt sympathy for Murrell, but the discovered outbreak changed that into a wild surge of hatred. Stewart had been driven out of the southern states as a lying rogue, and that harm could not be undone. The river people disciplined the Clan by laying by the heels all suspicious characters, flogging them severely, which modest reproof to the wild men was followed by brutal lynchings all along the lower river. So public opinion expressed itself.

Murrell the river god began to fail the minute he fell into the toils of the law. His friends dropped away. Stewart's accusations were established point by point. Like a half wild thing in a few days he became a different man, wound up hopelessly in a web of the court's weaving. His dare-devil courage was of no use in the prisoner's dock. His masculine willingness to hazard everything upon some open brutality of strength was hedged and harried at every point. The river god was caught within the cunning meshes of law and order. He was convicted and sentenced to fourteen years in the penitentiary. The judge thought him lucky, but Murrell knew differently. Fourteen years with never a horse between his knees, nor a sighting along the barrel of his fine rifled pistol, or a test of wits, man to man, with the loser to die? Fourteen years of law and order? Law rigidly enforced from morning to night, order, the first law of monotony? Oh, Murrell was true to his kind. Behind the bars there was no life for him. His eyes could never grow used to the short space of the cell and the walls that crushed him. He was fortunate in his era. Jails were dens of pestilence, filthy beyond thought. He caught prison fever under the close confinement. Perhaps it was just his luck. Who knows? Perhaps the Mississippi begged with Death to take its son, as wild and natural as itself, for he died in 1847, stubborn to the end, the last of the river gods.

CHAPTER TWELVE

PACKET PEOPLE

LIFE on the Mississippi steamboats has been called "an orgy of luxury." "Fast" and "elegant" were the terms applied to the boats in their advertising and there was something of truth in both words. Many travellers never lived so well at home as upon the boats.

Upon the steward fell the task of opening the day. He was the one officer on board who had no hours. He might be called upon in the night to subdue a madman, who knife in hand, undertook to make himself the only voyager to complete the passage. Gamblers, ladies-of-the-cabin, the negro orchestra, all made demands upon his time. There was little sleep for him and what sleep he got he must have taken standing up, naps caught hidden along the guard or in the shelter of some tall locker. His duties began in the morning. Before breakfast could be served, the sleepers who occupied the cabin floor had to be aroused and put on their feet. Up and out they went to pace the hurricane deck in the river mist. The savory odor of boiled beefsteaks and waffles sharpened their appetites, making them restless and impatient, long before their breakfast was ready. Rousing them was a dangerous task. The steward dared not leave it to a negro from the galley. These deck-sleepers were rough and men of prejudice.

The negro might have died suddenly, and he was needed by the cook. The steward had to undertake it himself. It was the beginning of his all too strenuous day.

Breakfast over, dinner faced him. There was no finesse to the cooking. A dozen lambs met the slaughter, or a dozen pigs, fit for the roaster, squealed out their protest and vanished into the starboard galley. The galley had two distinct divisions. On the starboard side the stewing and roasting were done in a cramped, odorous place where the heat was terrific. Unsightly cauldrons, red-hot broiling fires, business-like ovens, filled the place. Inspection of the starboard galley was never allowed. Perhaps it was clean, perhaps in its gloom and smother there was hidden many a layer of dirt; the control of those things fell upon the shoulders of the steward. In any case, passengers were never welcome there. If chicken were to be served they could see enough without prying into the mysteries of the galley. Chickens were carried alive in coops. A barrel of boiling water was made ready, a coop was placed beside it. The dresser drew a chicken from the coop by its head. With a clever movement the body was swung into the hot water, the neck across the iron hoop at the top of the barrel. The body was snatched from the scalding water, the head fell into the river. Other dressers stripped the feathers by two deft movements and a swift plucking of wing tips. They then passed to one who drew the larger pin feathers. An under-cook cleaned and cut up the bodies and, almost before the life was out, chicken dinner was being cooked. The lack of sanitation disturbed few, and if the spectacle addled a passenger's stomach, he did not have to look on. The port side of the boat was always more cheerful.

The port galley was the pride of the steward. There lived and worked his pastry cooks. Even ladies were welcomed in these quarters. They could look upon the art displayed with delight and understanding. The pastry cooks created delectable deserts: pies, puddings, tarts, creams, dressings, jellies. Their galley was a show place of the boat. Amid the aroma of freshly baked bread and biscuits passengers found nothing to shock their sensibilities. The pastry cooks were good men, proud of their work. At least once on every passage they prepared a true "orgy of luxury" fit for a noble palate. When the heavy dinner had been cleared away, then the realm of the pastry cook was spread before each passenger. His place at table was ringed by a delectable set of dainties. He might sample all, or any, at his pleasure, and each was a work of art. Custards, jellies, pies, all the galaxy of edible glory was his at a single sitting. A dozen dishes faced a man, a dozen kinds of joy. To the epicure it was a delicious moment, an instant of delight and dalliance. Which? All? That was the dilemma of the passenger and the delight of the steward's department.

Whether the voyager made much or little of his opportunity disturbed no one a wit. What was left went to the two score of cargo rustlers. To them it was served in no fragile glass. They never drank from long-

stemmed goblets, but in accord with their more primitive natures. It was not how they ate that interested
them, but rather how much they got to eat. The frag-
ments from the cabin table were gathered into flat pans.
Usually there were three sorts of pans. In one was
served the garbled staples: meats of all kinds, in picturesque confusion, mixed with freshly boiled potatoes.
In the second; bread and cake, and in the third; the
remnant of the pastry cook's delights. In some boats
these were specially treated to make them more palat-
able and less obviously mere left-overs, but usually the
route was direct from the cabin tables to the open
forward deck, where the men plunged in boldly enough,
each seeking what he wanted in the mixture. It was not
a pleasant thing to look upon this battle for food, not
a thing that promised much for the extension of civili-
zation to the roustabouts. The cry of "Grub-pile" was
an injunction to a sensitive passenger to seek the stern
of the boat, but the men seemed to get enough, to have
been reasonably healthy upon their mixed diet and
even contentedly happy. As on all the ships in the
world, there was no merrier moment in the day than
that brief respite of the meals.

The men and women of the passengers lived much
apart. The ladies took a turn on the hurricane deck or
sat along the rails in their chairs, but save at meal times
they saw very little of the men. The ladies' cabin was
their retreat, where most of their time was spent. For
the men, the bar was the centre. The life of the boat
was there. Free talk and boasting, an occasional wager,

a round of drinks, a quiet game, all these were to be found before the bar. The river boats never ran daily pools, nor regular betting on how many miles the boat might go in a day's run, as is done on ocean going steamers the world over. Occasionally, some one would offer a bet on the making, or failure to make a certain landing by some given hour, but it was a haphazard thing. The journey was too short to give interest in an organized pool, and through passengers were few. There was much short hauling on the river, passengers coming on board for a few hours or often only for one night.

Bar-tenders were sometimes very interesting men. Their position was one of difficult contacts. If a young man began his work with the enmity of a single officer his life could be made so difficult that he would move on in three months. Passengers were of all kinds, but along the Mississippi peculiarly free in their abuse if such attentions as they desired were not constantly forthcoming. They took umbrage at quite impossible things. A man wanted an egg-nog and demanded that the bar stand him treat. The bar-tender, being new, and not yet having the run of the trade, refused. The man went off, but finding an officer who disliked the bar-tender because he was a Spaniard, together they stole the crate of eggs from the bar and putting them into hot water boiled them all, returned them to the crate, and the crate to the bar. Then came the moment of triumph for the passenger. He asked the bar-tender's pardon. He wanted an egg-nog. He would pay for it surely.

There was no jest. The Spaniard broke the first egg from the crate with a flourish that nearly shattered the glass, and, as he realized something had gone astray with the egg, flung it through the window into the river. He grimaced at it, tried a second with like result, and a third. By that time he examined the eggs, found they had been boiled, and throwing tact to the winds he exploded in very satisfactory style, while the passenger asked for a cigar and strolled off. No, being a bar-tender was not an easy matter. Not the least difficulty was his relations with the gamblers who worked the boats.

Now there was never an honest gambler in this world, for the principle of gambling is to make a man think he is going to win while the gambler knows he is not. The usual gambling device was the playing card, and it was there the bar-tender and the gambler met. The bar sold playing cards. It was essential that a gambler get possession of several sets of these so that he could treat them in his own way. If a bar-tender yielded he ran the risk of being discharged. If he refused, the gambler had ways and to spare of making his life miserable, not to mention manhandling him when they met ashore.

Gamblers had several methods of treating these cards so that they could know their opponent's hands even from the moment of dealing. Taking the deck into the stateroom they were carefully opened. If the gambler "stripped" his cards, he withdrew from the pack the ace, king, queen, jack and ten. From the edges of these he cut with his razor a very thin sliver. This made the high cards ever so little narrower than the rest of the pack, enough, so that the gambler, whose business sharpened his sense of touch, could tell the high from the low in an instant. Other men marked them, carefully, minutely, in a way recognizable only to the keen professional eye. The cards were then returned to their wrappers, carefully sealed and given back to the bartender. When the day had advanced a little a few "sociable" games would begin. The gambler would take a hand. He would lose once or twice. Throwing down the cards he would call for a new deck. The bar-tender offered a sealed pack and for the second time the gambler would remove the wrapper, only this time he ripped it carelessly, confident the cards he held were his own helpful deck. If he were able to turn the sociable game into a deadly, earnest battle of wits against his sure knowledge, and the stakes ran to a few thousand dollars, he might remember the bar-tender by a hundred dollar bill.

Gamblers usually worked in pairs. Cleverly they boarded the boat at different landings and were introduced to each other as strangers about the card tables. Playing into one and other's schemes they broke large games into smaller ones, making certain that they gave their attention to those who had money, which they managed to take before the game was over. They could lose to each other with inimitable grace, knowing that a division of booty would come at last, if not on the boat, then later at some river port. By an understanding of some sort they confined their efforts to certain boats, avoiding interference with each other and seldom clashing in their efforts to lighten the money belts of travellers who sought a game of chance.

They were actors, able to assume disguise completely. On one journey one would play the part of a settler going out to take land, and the other of a pair would be a government agent. In turn they would become merchants, prospectors, or even lumber jacks, and on the southern river assume the regal manners of a gentleman planter. The negroes on the wharf would sing out as they boarded the steamer,

"Massa Walter, we're with you, sah. Doan go for to worry yo'self. We'll mind the plantation 'till you get back. Yes sah."

And the gambler, with a lordly wave of his hand would come on board, a sure enough planter in the eyes of all the passengers. A dollar invested in such a dusky claque was well spent. Sometimes the disguises were so well carried off that even the officers and the gambler's partner might be taken in. The art of the gambler was simple, once the play began, but the preparation for it was worth doing well. All suspicions were swept away. Cunning men!

In many streams there was a warning placard like this:

GENTLEMEN
playing cards for money
do so at their
OWN RISK.

This placard was often the gambler's starting point. Dressed as a planter he read it aloud in the presence of other gentlemen. He sniffed at its disparagement. He assured the world he never played for money, games of chance had no savour for him, no indeed. Then he waited for the spirit of the river to assert itself. The men of steamboat days were proud of one thing, their independence. They had come out of eastern cities to be free. What else had the river to offer them? They were men, yes sir, men. They were gentlemen able to decide such matters for themselves without the aid of the steamboat company. The placard was posted to relieve the boat of any responsibility, but some one was sure to be angered by it. The gambler had but to sniff in superior fashion, and some one would propose a sociable game. If the passengers were slow to wrath, the gambler's partner, dressed perchance, as a reputable medical man, would follow the opening and in a few minutes the cards would be called for at the bar and in the name of free and stalwart manhood, the gambler would be at work.

Superstition entered the play. If a man had a bad break of luck he believed it would be followed by two more before he could hope for a great deal. Another played very uncomfortably if there were a white horse on board. A white horse was very bad luck. Ill omens could be turned to good, if at a landing a red dog ran down, barking. By some a five-handed game was regarded as taboo and not worth playing in. Even with everything planned in their favor the gamblers had some belief in the gods of chance. It was after all a hazardous undertaking where anything might send the best of arrangments askew.

Social games allowed the gambler to pick and choose his victims for more serious work. In the early play he was able to see who had money, and who had none. Twenty dollar gold pieces were the most worth-while bait to draw him on, gold pieces carried in a body-belt which men never laid aside for a minute. Players showed their stripe in the early games. The man who would plunge doggedly on to ruin, the player, who, angered by an early loss, swept on in a high rage until broken, the fellow who needed to be encouraged by being allowed to win for a time, until confident in the smile of fortune he flung everything into one daring effort—and lost at the will of the gambler.

Once chosen, the victim was invited to a special game. He was a "sucker." This sounds like a term of vulgar reproach. It recalls the medium sized fish that rests so stupidly in still pools and will sometimes bite most unexpectedly at a bait, which he has often turned down. To the Mississippi gambler a "sucker" was merely a man stirred by interest so strongly that he neglected to be careful. Gamblers were often bored by the stupidity of these fellows, but they had a sport which delighted their souls, that of making a "sucker" out of another gambler, often a former partner. Knowing the man's passion for some game it was easy to make the most of it, see that he was beaten at it, and when he stood defeated in spite of his knowledge, to enjoy a good laugh at the fallen gentleman of the profession. The sad thing about the "sucker" was that he really believed he would win, believed it beyond anything his common sense whispered to him, beyond every shadow of probability at last. The gambler knew who would win. As he saw the victim's desire to devour him grow from hand to hand while they played, it is not surprising that he retaliated, and felt no remorse when the "sucker," having lost all he had, left the table broken, and beside himself with despair. Eat or be eaten was the law of the river. Eat or be eaten was the life of the gambler.

As he played at poker, or three card monte, the gambler saw fit to be courteous to men and chivalrous to women. He depended upon their good will for play. His life gave him an ease of manner among the motley crowd of the boats. Often he was a considerate man, generous to beggars ashore, returning bits of his winnings when the loser proved game and had been made needy by his play, making friends by consideration and even homely kindness. A poker game was to a certain degree, always social, but when it came to three card monte the gambler had to be obviously determined as well as coolly self possessed. Poker we still have with us, but monte is seldom seen, scarcely oftener than the sure thing of the thimbles and the pea which used to feature circus days when there was much of fraud associated with the side shows. There were several variations but the usual method was to take three cards, one of which was a jack. These were shown to the player

265

so that he thought he knew which was the jack. They were shuffled then, and the gambler made a bet that the player could not pick out the jack from the three. The player, full of confidence, bet, and failed to turn up the jack. The gambler pocketed the stakes. Some gamblers introduced this game as a card trick, built up interest in it, got passengers to bet in a small way. Interest grew. Then, when they were well heated, the gambler would cry out to his partner. "I'm betting five hundred dollars you cannot turn the jack." The partner, his partnership unknown to the passengers, counted out five hundred dollars. The gambler shuffled the three cards, spread them face down. The partner turned up the jack, took the money and walked out, while the spectators gasped. The gambler, weighing his audience, knew that if there were any money in the crowd he would have it. Some one would ask him what he would bet. The gambler then would size up his man, put the amount as high as possible, and at the drawing the jack would not be turned up. Another would chance it, and another. Even if the game were played honestly, the gambler had two chances to the player's one. Some men palmed the jack in shuffling and never put it on the table at all. Some bent the corner of the jack and in the shuffling substituted a card with a similarly bent corner which, needless to say, was not a jack. This puzzled the player, but curiously enough made him more eager to play again. He had failed although his eyes had seen the bent corner. He could not understand it.

So long as players remained gentlemen, pocketed

SCARCELY A GAMBLER EVER WORKED THE RIVER WHO DID NOT HAVE TO
FIGHT

their losses like good sportsmen, and did not stop playing too soon a gambler's life was easy and pleasant.
That, however, was more than could be usually expected. Scarcely a gambler ever worked the river who
did not have to both run and fight at times. If the alter-
cation were over money at the tables of play a display
of an adequate pistol worked wonders to keep anxious
hands from scooping in disputed earnings. Gamblers
never had much use for anything they could not put in
their pockets. They might win anything: negroes, dogs,
chickens, cows, but they translated any of them into
money at the first opportunity that offered. Travelling
in the height of luxury they could not go around armed
with a holster pistol. It would have frightened the
ladies and made the gambler's business in travel too
conspicuous. The fine rifled pistol of Murrell was not
what was needed. A gambler seldom used a pistol to
shoot at a greater range than ten feet. He needed a
heavy calibre weapon, one that impressed a man with
its serious attitude toward life, as he stared down the
black throat of its barrel. For the gambler's service was
invented the derringer. This was a short, double-
barreled pistol. It fired a heavy slug of a bullet from
its rim fire copper cartridge. The three-inch barrels
were too short to give any base for sights. The barrels
were placed one above the other instead of side by side
as is usual in double-barreled weapons. The bore was
forty-one calibre, well on its way to a half inch in diameter. Thrust at one across a pile of money, which had
tempted eager hands to seize it, the effect of a derringer

was tonic. It stiffened the moral fibre of a man about to turn bad loser. On more serious occasions it defended the gambler's life from the murderous attack of some player whose losses had turned his head. A glance at a derringer's ugly snout had a tendency to check an uplifted knife in mid-air, or to make a hate-flushed face turn ashy white. The derringer was a good pistol but it had only two shots. For that reason the advent of Colonel Colt's new patented revolver, in a compact form, was very welcome. The barrel of these weapons was longer, and since they lent themselves to rapid shooting, the six shots were hailed as a great improvement. Many gamblers felt sentiment about their pistols. They named them the "Betty Mary" or "Jonathan" or became historical and called them "Andrew Jackson."

When it is considered how many gamblers worked the boats it is surprising how few shootings there were. It was as if the knowledge that any man might be armed kept men from thinking lightly of killing another. Fights there were daily, brutal, unsparing struggles, but fights that frowned on knives, or blackjacks, or the whipping out of pistols. There was no place for such weapons where bad blood was leaping. They were reserved to defend one's life, to guarantee one's property, or to stop the rush of a crowd until some one of the many could be invited to a fair fight in which there were no weapons but a man's hands.

Occasions rose in which gamblers were unable to trust to their pistols. Several hill men from Kentucky, or the Carolinas, would get together, decide they had

He stared down the black throat of its barrel

been wronged by the gambler, and going below for their own revolvers would begin a gambler hunt. This was peculiarly liable to happen when the losses had been heavy, and the gambler had cleaned out the money belts of all on board.

When the search began the gambler had to leave the boat. If he left in plain sight he had an excellent chance of being riddled by bullets at the gangway. A partner, unknown to the passengers, was a great help in such a matter. The winnings could be turned over to him with an understanding as to when a meeting could be made later. Having nothing else to look to then but his life, he was free to use his art. Sometimes at the landing, an itinerant preacher would push his way through the crowd, and apparently contemplating heavenly things, would walk serenely ashore, past a half dozen or more angry men waiting at the gangway. It was of course the gambler in disguise, turning a last trick upon his victims. It was a tense moment as he walked on to the wharf offering his undefended back to the bullets of his enemies, and one that chilled the blood, but a gambler had to be equal to it. Brazen courage was part of his equipment, and a good disguise scarcely more necessary.

If the victims were peculiarly awake, and so daring a landing promised failure, the gambler plunged among the negro roustabouts, blackened himself, purchased a deck hand's clothes and turned to at unloading the boat at the landing. Every passenger came in for the closest scrutiny but no one heeded a negro staggering ashore under a load. Once on the landing, the gambler dropped his burden and made off. Sometimes his flight was observed and a rain of bullets followed him, but the surprise of his detection was apt to make the aim inaccurate and he usually escaped.

The worst thing that could happen to him was to have his pursuers begin the hunt on a stretch of river where the distance between landings was very great. The odds in favor of the hunters were then increased many times, unless the gambler could persuade the pilot to cut close to a sand bar. This the pilot was usually willing to do. The gambler had to choose his moment then, hope that the water was not too deep for him and make a long leap for safety. He disliked to do it because it was a confession his wits were not equal to those of the angry mob. Even if he escaped he was very funny to the whole boat when he came up wet and draggled, found his feet and started the long wade ashore. The pilot usually pulled the boat sharply back into the stream, swinging the stern so that even a man watching for a shot was likely to be confused. If the gambler were clever in the water he remained below the surface as long as he could to defeat the flight of a stray bullet. Once the long moment was over he used his knowledge of the country to keep his pursuers from coming up with him. It was an arduous life, but one that appealed to a vigorous, resourceful man.

The pursuit was sometimes carried on resolutely. A gambler who boarded a train might find he had been trailed. Sometimes his winnings would be well into the thousands of dollars: it was not surprising his victim did not quickly forget. He might find it necessary to say a prayer for himself and step off into space from a rapidly moving railway coach. He might find himself in a hotel looking into a pistol in the hands of an angry woman whose husband had been "plucked." His legal status was nil. There was no protection for him but in his own efforts, yet few gamblers remained rich men in spite of their earnings. Money they gained at hazard of their lives they played again in the gambling houses of New Orleans or Natchez. Play was their bread and butter, and play was in their blood. Practically all of them had some game of chance in which they ceased to be the master and became that excited, hopeful thing, the "sucker." Always they hoped to break the bank. Always they looked for the thrill to be gotten out of a game where not all the factors were known to them, a game of chance run by other men for profit. Man learns so little in one short life, and his love for uncertainty is very great. Gamblers really gambling are a sad spectacle.

The best story of a gambler setting his pursuers at naught is told of a southerner. He had cleaned up about five thousand dollars by the old, time-tried three card monte. There is something tantalizing about losing money at three card monte, something that sets men's teeth on edge. It seems so simple to be able to turn up the right card out of three, and yet, thanks to the gambler, it is seldom done; never when he wills otherwise. Five thousand dollars was a winning. Most men in the game were content with a few hundred as a week's earnings, but there was nothing of the piker about the southerner—five thousand, no less. The victims went for their pistols and then for the gambler.

Now the normal thing for the man who ran the game to do was to leave the boat by any one of a dozen stratagems, but this man was cold as ice, and besides he wanted to reach New Orleans. Perhaps he had an engagement there, or perhaps he was eager to plunge his winnings against the bank of some gambling house. The story does not state. He decided to stick it out and stand by the boat.

He was fortunate in that he had been raised on a plantation. The negro speech was natural to him. Taking his courage in his hands he used a little grease paint, changed his clothes and became a member of the boat orchestra. There were eight men in the orchestra. They played the violin and guitar. He played a banjo. The real negroes of the orchestra passed their day as waiters, as baggage hustlers, as deck hands. The gambler spent his trying to act like them and succeeding. His pursuers searched the boat from pilot house to keel. They located his cabin and found no one there, and the bed made up. They prowled about the hold, the galleys, the forward deck. They met him twenty times a day but never knew it.

He passed the hat for the orchestra, he played a good banjo. At night he slept in the sounding-boat swung from its davits on the hurricane deck. Nightly he played, played for dances in the cabin and when the boat made landings he sat with the negroes on the guards thwanging away on his banjo, letting out real African guffaws, working hard to attract trade to the boat. His orchestra was good, one of the best on the river, but his playing was equal to it and his grease paint held. The passengers never thought of the tall, thin negro with big feet and sprawling hands that made the banjo talk, as the dapper customer who had relieved them of their money. He heard them talk on the hurricane deck as he lay under the stars in the sounding-boat, heard them promise the gambler a thousand awkward and disturbing deaths, but he played his part, determined to reach New Orleans.

A cautious man would have played them safely, but not the gambler. He had a good voice. He knew the negro patois to an accent. Boat orchestras sang in the inimitable, soothing manner of the negro, sang to a swing of the body and the merry bobbing of heads. The gambler's wigged head bobbed as happily as any. He sang a song or two he had learned in childhood, simple, plaintive tunes taught him by a plantation mammy. The orchestra caught his spirit and sang with him. It was a happy trip. The steady plowing of the boat took him toward New Orleans, toward safety with his five thousand. His victims still looked for him. Cold as steel, with his pistol always at hand, he faced them night after night, his tenseness never entering his singing, his banjo defying fate that had been so kind to him, to do its worst.

"Fortune will play its part, let man do what he may."
Fortune stood by him.

The boat made the wharf. The victims, all armed, watched the passengers land, looking for their man, suspicious of every beribboned, full skirted woman who went ashore, but there were none tall enough to be the gambler. He passed them and returned, hustling baggage, unloading freight, hard at work and talking all the time. At last they gave it up. The gambler was not a man to throw away a winning by ill advised precipitancy. He stood by the boat until the small hours of the morning. Then he went ashore, a white man again. A lone man sprang at him from behind a pile of bales. A knife flashed, but the gambler was ready. His pistol barked. The fellow, looking surprised, toppled over and fell from the levee into the water. The gambler went his way, his winnings safe in his sack. His daring adventure was ended.

The steamboat orchestras became an institution. A good one drew trade for it was the life of the boat. It was hired to play at the will of the captain or the chief clerk. In addition to their pay they were allowed to pass the hat and keep what they took in. The business side of the agreement began and ended there. For the rest, the orchestra aimed to please. It played or sang at will. At the cabin dances they gave their best. Usually there emerged from the six or eight some primitive spirit able to rule. When the boat lay along the levee he would lead his little troupe out on to the guards. With a native genius for leadership he would take his stand under the flickering glare of the torches and sing.

THE STEAMBOAT ORCHESTRA BECAME AN INSTITUTION

"Ah fires all day at the middle do'
Ah sings all night fo' yo' folks asho'!
Shake 'em up libely for to make de boat go."

It was a creative effort, simple, barbaric, but undertaken with a swagger and a gusto that made up for much. To look at anything was to get the impression it gave. All was inspiration. A glance at the captain standing on his hurricane, eager to cast off and get upriver, won a tribute that nowhere was such a gentleman to be met as himself. The engineer was praised for his daring in carrying steam, the pilot for his stubbornness in demanding and getting his right of way, the chamber maid Chloe for her dusky loveliness. Last but not least, the singer sang his own praises which left no doubt as to his importance.

Such a man might at heart be a good fireman, but he was too important to the boat to work at it much. Instead he acted the part of general foreman for the negroes on board. He had a gift; could he not roar out in a curious negro minor,

"De big wheel rolls, her head comes round,
We're bound to go, 'less we take de ground."

or if the roustabouts needed a bit of jollying to help them trundle freight more smartly.

"De debbil he come in the middle of de night
An' he couldn't do nothing 'cause the boat was tight
But I was scared most nearly white
Ah haah, de levee!"

The other six would strum furiously then and join in by way of a refrain,

"Ah haah, de levee!"

Jocose stuff this, poor in meter, but alive. It was the play for a word of praise, a laugh from the passengers. Behind it lay that fierce thirst of a man to speak his piece under the dispassionate sky dome, to stamp something of himself upon the great river before he was snuffed out by time. It was of life, human, opposed to the tireless might of the broad waters. If it were unnatural, and forced, the more credit to the creator of it, ignorant, uncouth, reaching out after the joys of the gods.

There were times when the steamboat clipped along across a white moon-path, slipping down-stream easily, when there was time for reverie, and then the orchestra would be itself for a brief spell before the dance began. Back over the years they would be carried, beyond all the days they had known, back to a world of fighting black men moving swiftly like shadows. Muffled, and almost lost, the beat of the tom tom stirred their blood out of forgotten things, things African, filled with the lush, damp tropics. Like a moan of lament the soft minor chords blended. Tears came to the eyes. Rough rivermen thought with regret of the childhood they might have had. Children listened, dreamy eyed, learning in their hearts of life they had not yet lived.

"De night is dark, de day is long
And we are far from home
Weep, my brudders, weep."

Simple words of the river, of all rivers in strange
lands. Sorrow and homesickness, perhaps, but the
singers on the boat had never known a home half so
good as their berths down forward. No, it wasn't either
of those things, it was just a plaintive plantation melody,
and they were carried off into the moonlight by the
crooning lilt of it at work in their hearts. It carried off
all on board; the pilot, the engineer, the gamblers, the
women aft, and the boat slipped down the great river,
slid on and on while the negro singing rose and fell.
They had forgotten their instruments then. Neither the
guitar nor the banjo had voice. Ahead lay the lights of
New Orleans flung in a great crescent, a lantern bobbed
here and there close to the water, and the voices of the
negroes sang boldly,

> "De night is past, de long day done,
> And we are going home.
> Shout, my brudders, shout!"

CHAPTER THIRTEEN

MIGHTY MASTER OF ARMIES

THE Mississippi remained neutral in the Civil War. Its steamboats left their regular business for the most part, and became gunboats or rams. Its men took sides, joined the army of their choice, or the navy, and changed both their subjects of conversation and their habits of life. While they were engaged in deadly struggle the great river paid no heed. If a stream could jest it often seemed to mock the efforts of the warriors. Let a breast-work or a bayou fort need but four hours to be finished and the water was bound to rise and wash the incomplete defense down into its silt. If a gunboat chanced a shallow climbing there would follow a drought and low water to prevent its rejoining its companions. The attentions of the river were not strongly partisan. It was no respecter of causes. Any of the various hazards it presented were only its age-old tricks, ingenious enough to baffle the human mind at any time. The current trifled with the intentions of both sides, preventing the control of gun fire, throwing maneuvers into confusion. The headstrong, heedless strength of the river seemed capriciously opposed to the war. Yet both sides frequently reached success. The malice of the river was like that, it was even so impartial. Perhaps, after all it smiled.

Anyone would have smiled at the gunboats which took the place of the gilt and crystal fliers. They had paddle wheels housed in their sides, sides made of iron plates bolted to crib work of heavy timbers. The sides sloped at the top, all four of them, and the guards were widened so the overhang sheltered the paddle blades. Underway they looked like an old-fashioned, black, iron bread pan afloat upside down, or a block house gone adrift. Behind the shutters of iron plate were the guns, which, with the added load of crew and provisions made the boats unwieldy craft. Some had iron beaks for ramming an enemy. Once they went into action they ceased to be a laughing matter and became grimly murderous, but even shrouded in smoke clouds, stabbed by the flash of fire from the throats of their guns, they were a far cry from their sister ships of peace, the fliers who boasted speed and elegance. A rumor of the approach of the iron clad craft struck fear into a river town if they were enemy vessels, and genial respect if they were friendly. There was no denying they were business-like craft.

The cities of the Mississippi were as unconcerned as the river itself in the early stages of the war. Their citizens changed their minds from time to time. Some were now northern, now southern in sympathy, at the turn of a wheel. In St. Louis, for instance, an order forbidding the flying of the southern flag was issued. A young southerner, forgetting that in war times tongues must not wag, declaimed against the order as an intolerant thing. He waved his arms in gesticulation as he rode along in a street car behind a pair of laboring horses. He talked to everyone, everyone, that is, but a little man in the corner. Anyone could see this listener was nobody at all. He was short, shabbily dressed. His efforts as a farmer had brought him failure, in business he had succeeded no better. That was the reason he was bumping along in the flat wheeled car. He was bound back to his father's leather store in Galena to clerk out a monotonous livelihood. There was worry written in his face, he had the grey skin of fatigued middle age. His hand was unsteady. At thirty-nine hope is dim and drink has been many a man's way of making it glow again if only temporarily.

The young southerner ignored the little man in the corner, but the steady eyes met those of the raving youngster doggedly. Then he spoke in a tired, quiet voice that rumbled a little hoarsely. The lad listened amazed. Heads poked out along the aisle to see who had taken up the challenge. The words were few. The little man said merely that perhaps the intolerance was not so great after all. Even though there had been much flying of the rebel flags of late, he had noticed so far that no one had been hanged for it, although many deserved to be. When he finished he met the southerner's eyes fairly, and the supporter of tolerance saw something there that he could not but admire. There was loyalty in those eyes. The little man was a son of the great river. His boyhood had been passed there. In his failure he was going back to it. He had fought for the northern flag in 1845 against Mexico. He might be little, but he was not afraid, and he spoke what he felt to be the truth.

Within a year that man was colonel of an Illinois regiment. In 1862 he captured Fort Donelson as a general. He stuck by his valley, keeping up the work his lagging spirit had found to undertake on the St. Louis street car, fighting for possession of the great river. He was at Island No. 10 and at last he was set to face the great southern stronghold, Vicksburg. The little man was U. S. Grant.

Vicksburg was a natural river stronghold. The northern forces had unlocked the Mississippi by 1863. They had passed the barrier of New Orleans. The taking of Fort Henry and Fort Donelson had preceded the reduction of Island No. 10, but Vicksburg was left in the hands of the southerners. Sitting atop a high, clay bluff, with the river at its feet the place was a citadel. Pemberton, the confederate leader, had made it a stronghold. Military men said it was as strong as Sebastopol, nay, as Gibraltar itself. It bristled with guns, batteries that swept the river both north and south for miles. From the northeast the Yazoo River drained into the Great River almost beside the town. To the east and south, bayou, swamp and forest protected it, lowland that was dominated by the great bluff that reared its heights more than two hundred feet. To aid the batteries there were troops, willing and able to check a northern thrust through the bottom land, and rams and gunboats on the river to meet the federal fleet. Twenty-seven thousand troops garrisoned the city. Three thousand inhabitants stood by their houses. U. S. Grant undertook the attack.

He threw his army about the fortress, cutting it off from the world, and began his siege. The inhabitants lived a nightmare. The federal guns filled the streets with iron debris. The non-combatants dug caves along the river front, high up in the clay wall. When the cannonade opened they left their houses and lived in the galleries they had driven. There were so many of these rabbit holes cut in the clay that the river face of the bluff looked like Gibraltar at close inspection, only instead of guns built into the protections whole families lived huddled in the burrows. Sometimes a chance shell filled the mouth of a hole with dirt. Then the inhabitants had to dig their way out quickly or suffocate. They soon learned to have two openings to their underground homes. If one were closed they escaped by the other. Even with the honeycombing of the bluff there were never enough hiding places. Caves able to hold eight or nine had to shelter sixteen. The air was deadly. Candles would not burn in it. Food was short. Fright sickened many. Misery was everywhere. Many lived but a few years after the siege and all were sadly shaken and battered, so that they were never themselves again, but became sad, distraught creatures, jumping at their shadows, permanently broken by the experience.

No news came to the city. There was no trade with the outside world, no running down cheerily to greet a steamboat at the landing, no clatter and roar of departing trains. All normal life closed and in place of it began a furtive existence, fear-shadowed and horrible to contemplate, but much worse to endure. Flour sold at two hundred dollars a barrel and bacon at five dollars a pound. Fresh meat was unheard of save for the slaughtering of a few mules; green food, a thing seen only in a heart-breaking nightmare. In the streets was the silence of despair, the deathly silence of war. Those who passed moved soundlessly with frightened eyes on the heavens, waiting to see the first shells of the renewed bombardment. At irregular intervals the guns would shatter the air, tear at the ear drums and the chaos would set the earth a tremble. Half crazed, the inhabitants would dive for shelter, and the soldiery, who dared not run would mock them. "Into the earth, ground hogs. Fly rats, dig with your chisel teeth. The Yankees are hunting for you." It was their way of preserving sanity under the baleful strain of war. The taunt gave them courage to continue.

The iron hail was terrific, the bursting of countless shells like nothing but the end of the world. Houses were honeycombed from side to side. There was not a whole window pane left in the town. At the setting of the sun the terror and agony was greater. Horrible groans followed some sadly fortunate shot. Shadows moved furtively against the lurid glare lighting the sky to the crash and shock of great guns. The spitting drone of musketry was as silence to the bestial clamor. A shell exploding revealed the frightened faces of women and children, half distraught, faces aging by

minutes instead of years. Daring residents gathered un-
exploded shells and, under cover of the dark, set them
in what had been their gardens. Souvenirs they were,
and their collectors were often dead before the sun
shone upon the ridiculous mementoes. Sometimes there
would be a blinding flash, a wild scream of pain. A
flying shell had struck some one carrying a "souvenir"
and both had exploded. Savage nights in old Vicksburg!

The gunboats engaged upon the west. General Sherman tested out the strength of the city by an assault
and was flung back in jarring fashion. Grant had twenty
thousand soldiers at his command but he dared not
fling them into a smothering attack. Vicksburg was still
lord of the Mississippi. The strongest batteries faced
the north, holding Grant's army in check. He therefore sought to get his men below the city. Behind the
city, and to the south of it was an endless network of
shoal creeks, lakes, and back waters. Grant selected a
point where the turn of the river helped and put more
than a thousand negroes to work shovelling their way
behind the stronghold. They worked willingly for
several weeks cutting a canal across the peninsula. It
was heavy work. When it was finished the gunboats
and troops would come down. The negroes toiled and
sang, bantered and suffered, but they could not know
what was to happen. There was caprice in it. A deluge
of rain fell. The Mississippi, envious of human powers
attempting to interfere with its course, or filled with
the wild will for madness, or only swollen with heavy

rain, if you will, gutted the unfinished canal. Breaking its banks, the muddy torrent tore and sucked at the shoring, caved in side walls, washed the work away and left a morass of wasted effort.

The next attempt was to see the river turned directly to use. It was natural that Grant, as a river dweller should think of its possible service to his cause, but it was Porter, of the navy, who undertook the affair. He knew there was a network of water lanes behind the towering bluff. Was it possible that that network knitted so well that gunboats could be worked through it and brought out into the Mississippi below the city? A man was sent off to explore. In a steam launch he found a way through by a route which led first to the north, up the Yazoo River, then by a wide bayou into an intricate stretch of forest travel leading out to the Mississippi again. He believed there was water enough to float the fleet, but there would be need of good axemen to clear a way through the forest, and tugs to tow off any gunboat that went adrift. Then Porter decided quickly. The tugs and axemen were ordered up. The gunboats were put at it and the transports were given orders to follow through. The soldiers were marched down the west side of the river, out of harm's way. If the transports got through they could ferry them across and a flank attack on Vicksburg could be undertaken.

General Grant reviewed the plan but added an idea that a riverman would think of in a moment. They would first cut the levee, letting the Mississippi fill the waterways to the full. Porter retired on board his gunboats and, in the first rush of the river let loose, plunged upon as queer a journey as ever gunboat made. Tree trunks checked the boats. They were thrust from side to side by firmly rooted stumps. Overhanging branches swept away the standing rigging of the signal masts and tore at deck houses or wiped their length across funnels, the guys of which quivered, taut as a fiddle string. It was hard on the boats, but it was worse on the axemen. The forest was always there to be fought with. One embrace loosened, another waited five yards on. They had to cut away cleanly and keep cutting. There was an element of time in the attempt. Sooner or later their advance was sure to be discovered by confederate scouts. A genuine delay, checking them for a few hours, would prove serious, but meanwhile, until the confederate opposition became real, there was nothing to hold them up except the tiring of the axemen, and they at last cut their way into a seeming river, free of trees and two steamers broad. Cheerily they steamed along. Then at a bend they came suddenly upon a newly-built fort set squarely across their path blocking the waterway. They backed water in haste, crowding and piling into one another. The first boat opened fire and the new fort tried to reply, but it had been finished only within the hour, so there had as yet been no ammunition served to the guns. The gunners tugged at useless lanyards, forgetting their plight. A flank movement of troops would have found the gunboats in sorry sort. There was no sensible movement for Porter but a retreat. This the boats began at once with engines in reverse since there was no room to turn around, making their way back to the fleet waiting above Vicksburg.

But Porter was not finished. He found another lead. He followed it with four of his big gunboats and two mortar-boats. This route took him much farther from Vicksburg. Gloomy swamps, tall overhanging trees, flanked the new route. Sherman worked with him this time, travelling a parallel course and keeping as close as the nature of the country would let him march his men. If the other trip had been unusual this was eerie in its way. They travelled for days, ditch crawling. The forest took on a grandeur, shutting out the sky. Porter's men were seamen, used to salt water, free wind, a rolling deck and a far horizon to greet the eye. They disliked the strange sights and sounds. Raccoons, wild cats, and occasional bear disappeared from before them. Small animals pattered about the deck at night. The bayou was narrow and belonged to the forest. Sycamore and tupelo gum-trees stood sullenly along the bank. The gnarled live oaks glowered darkly upon them, old as the hills, old as the gloom of the backwaters. If they found a clearing planted in rice or corn they could be sure that just beyond waited a noisesome cypress brake, a swamp realm where the waters were black and foul and the trees were hung with Spanish moss.

The going got worse and worse and Porter began to fear the Confederates would get wind of his maneuver and front him with another fort in some narrow pass. The trees tore away life-boats from the upper deck and downed a smoke stack or two. He said he needed no boats, only his guns and engines. When a tree blocked the narrow way he put on steam and rammed it. Soon he found himself in a pinching gulley. His boats were forty-two feet wide. Cypress Bayou was just forty-six. Moreover, the levee banks of the canal rose above the gunboat decks, making their guns useless since they were left so low they could only shoot into the canal wall.

It was then they came upon the natives of the district. The open country was raising everything it could to feed Vicksburg. What little the townsfolk got came into them from this section. On all sides as they advanced the civilians retired carrying off their possessions in wagons, carts or barrows. What they could not take they destroyed. At one place on the bayou six thousand bales of cotton lined both banks, waiting shipment at some fortunate turn of the war. Porter's gunboats stirred up their usual panic and the fleeing natives set fire to the bales. A dense mass of smoke hid the bayou, cut off the sky. The wheelman stood steady but looked inquisitively at Porter. Porter decided the matter in a flash. A mass of cotton like that would burn a day or more. He had seen the docks at New Orleans when the townspeople had set their bales afire. It was hard to navigate the bayou in any event, but he could not afford to wait.

"Full speed ahead, pilot, and mind your eye," he said.

The pilot rang the bell once and looked at Porter no more.

Headlong into the smoke and flames the flotilla passed. There was a two-knot current to buck, and the gunboats were not fast at the best. The iron plates lost their paint in great blisters. The ports were all closed, the crews standing at their posts for fire and not to serve the guns. On deck the heat was unendurable, below it was sweltering. The pilot wrapped his head in a cloth. Hoses were turned upon the decks. In the midst of the flames and smoke they held on. Porter's boat shook like a stricken thing, came to a dead stop and then lunged ahead. He had carried away fifty feet of a bridge at a blow of the ram. On he pushed, determined at all costs to get through. He sent a tug ahead. The tug came upon a green scum on top of the water but pushed past. It was a mass of growing basket willows. They fastened upon the hull, the shoots dragging along the bottom . She did not get fifty yards before her headway was gone. Her wheel could not move her ahead nor astern. Porter flung a gunboat into the tender, green shoots, but she could not even reach the tug. Frantically, under the urge of their officers, the crew, with hooks, saws, sharp cutlasses and ropes tried to get free. Haste was the watch word, haste and forward.

As if to verify the wisdom of Porter in making all possible speed, first came reports of the arrival of Confederate troops and almost immediately shells began to fall on the gunboats. Porter counted the seconds between the arrival of the shells and the noise of the cannon that fired them, guessed at the direction and setting his mortars to reach the suspected spot silenced a battery he never saw. He gained a little time to think, but he wanted that in peace. This gunboating in a canal was worse than he had thought it.

Why did not Sherman come up to cover his flank? He sent a telegram to Sherman by a negro who claimed to be the "livest kind of telegraph wire" and who made no bones of the fact he would carry word to Sherman or to kingdom come for half a dollar. Porter wondered if these were one and the same place, and sent the negro off pell mell. Porter was badly beset. The enemy were all about him. The willows had his squadron fast. Any advance would be contested every foot of the way, for the southerners knew the menace of that ugly little fleet of iron-clads. After he had stopped the fire of their batteries he heard nothing from them. This added to his worries.

It was then the Great River played favorites. The water rose suddenly. The boats were freed from the willow tangles. With the rising the current increased, so that to go ahead Porter would have had to travel on between high banks even more slowly than before. He took the omen of the river. By profession he was a salt water sailor, a man out of the frigates, loving well the wind in the rigging and the swift flight of the smooth, deep water hull. But to a man who watched and read wind and weather at sea the rise of a stream was enough. He would retreat. His boats could not be carried further safely into an aroused country. His fleet slid back to the north, wallowing a way through the ditch.

The current was with them now for the bayou they were following emptied above Vicksburg. Their speed was good. It was risky to charge on blindly around bends at the rate they were travelling, but Porter was anxious since he could not work through to success to make his retreat while the enemy was still surprised at his mere presence. True, he had sent to Sherman. He had seen the negro pack the message in the kinks of his black wool, but had he gotten through? Had he played fair and not merely turned his message over to the southerners? Porter had reason for anxiety. He was in a very tight place, where the slightest misfortune might turn into complete disaster.

As they twisted and dodged along the narrow channel the gunboats suddenly came upon two trees dropped into the stream, blocking the way. This was serious. Could they fight their passage back through those days of weary travel? Even as they struggled with the first two logs the sound of the Confederate axes rang through the forest, felling new obstacles ahead. Porter felt he could work blindly no longer. He sent out men to act as eyes for the expedition, and learned that ahead were gangs of negro axemen felling trees at a great speed. It was time Porter should show them how the navy did things. He had tackles rigged, using as large blocks as he had and a long fall (the rope that ran through the blocks). Then he steamed on to the next obstruction. Several seamen were landed. One climbed a tree taking one end of the tackle with him. One took the other block out on the freshly felled tree and made it fast by a strop about the trunk. The end of the fall, was made fast in the iron-clad. When all was set, Porter, watching from the deck of his gunboat, rang two bells for slow astern and then when the tackle was taut rang a jingle. The paddle wheels beat the water. The fallen log's end rose toward the tall tree where, high on the trunk, was fastened the upper block of the tackle. When it was swung clear, Porter stopped his boat, came ahead, lowering the log beside the canal and when both blocks of the tackle had been loosed in readiness for the next obstacle, the fleet went upon its strange voyage. It was sailor's work, not so bad on the backs of the seamen as sending down spars or striking a topmast on a square-rigger. The steamboat did the pulling which made the work easy. The crew were smart in their part of making fast and letting go. It was welcome sea-going work after days of puddling about ditches. Soon they gained speed, working like a gun-crew to cut off seconds at each process, studying how this strop was best passed, how to clap the tackle over the bark of the forest giants, how to take the slack out of the fall with the greatest despatch. It was not long before they were destroying the obstacles faster than the axemen could supply them. At last they were up with the latest felled tree. Then the tug pushed on, to prevent by howitzer fire any further checks being given them. It looked as if the gunboats would win free, but the enemy had still to be reckoned with.

The men Porter had put ashore to establish contact with the Confederates brought in two commissioned

THE CONFEDERATE ARTILLERY CAME AND SHELLED THEM

officers as prisoners. He learned from them he was surrounded, that Pemberton, in command at Vicksburg, knew all about his movements but did not seem to know that Sherman was working with him. To all appearances Sherman was no longer concerned in the six lonely iron-clads. The Confederate artillery came up and shelled them. Sharpshooters, posted in trees, tried their hand at clearing the decks. Lieutenant Wells was killed beside Porter. The officer who went to his assistance fell in his tracks. The action was growing warm. Standing in the openness of his boat, Porter saw a long column closing in on him, picking its way through the woods and avoiding the fire of his mortar-boats which were shelling the forest. So it had come then, he had lost his boats. Snowed under by the number of the boarders his men would be unable to keep them. He turned to give orders to do what they could when he saw the column in grey falter, break their formation and actually wilt away before his eyes while another in blue took its place. The negro had gotten through, Sherman had not been lost, he was very much alive indeed, as the Confederates moved off from before his vanguard. For them, Vicksburg was a defensive action; they had no men to waste exchanging blows with Sherman, but they were chagrined to see the gunboats, they had felt so sure of taking, slip free.

When Sherman rode up to see Porter he merely said;
"Your gunboats are enough to scare a crow."

Porter had not thought of that before, but it was true. The flotilla had endured a hideous battering.

Davits, stanchions, boats, everything that projected above the deck had been torn away by overhanging limbs. The funnels, deck houses and companionways were splintered and bashed by the heavy rifle balls of the sharp shooters. The shells had dented the armor plates, here and there, opening cracks in the iron. The blazing cotton had scorched and blistered the paint so it hung in ragged ribbons. If the Confederates had used solid shot on the plates of the gunboats they might have cracked them wide open, or at the very least strained the oaken backings, so that the hulls would have opened and the boats filled faster than the pumps could have handled the water. The gunboats in turn used grape shot where the gunboat guns would serve and shell only against the batteries. Everything about the boats was worn to a shadow of itself. They had done things never undertaken before or since; a queer, outlandish adventure.

When the last boat floated clear, anchored in the Mississippi, Porter had his first free moment before he undertook the repair of his fleet. For the first time he went back over the adventure and could find no fault in it, but there was a strange discontent in him just the same.

He talked in friendly fashion with his officers. He went over the latest reports from the commanding officers of the other boats that had gone with him where steam had never before driven a boat. For a few minutes he struggled to compare them, to examine various versions of some happening in which he had wondered at the actions of his subordinates, wondered why they acted as they did, and what thought lay behind the effort. Of that too, he tired almost at once. He could not account for his restlessness. A naval officer is trained to long vigils where patience is vital to his peace of mind, and Porter was as fine an officer as the navy ever had. A man at the peak of his profession should have had more self-control. So thought Porter and dismissed his lack of ease into limbo, but it would not go.

Climbing to the steel deck he strode fore and aft, as was the custom of a sailor, not thwartship as the steamer man does today. His heels clicked upon the metal. He was ashamed of himself, but he walked to put himself at peace, and when one walks with such a purpose one turns and turns again without a thought. A lethargy stole over him as he looked down on the greyish water, and he began to hum to himself a queer little sequence of sounds, not a song, not a hymn, but rather, over and over, half a dozen muffled tones just audible to himself. His humming was timed to his pace. Gradually he gave himself to the lilt of the thing. He felt at ease. Of a sudden he halted and smiled. He knew what that humming meant, he knew why he walked, he knew why both eased his restlessness.

They had taken the place of the churn and beat of his paddle wheels. For days and days he had heard them dashing the water up into their paddle boxes; a pleasant sound, and at every turn of the crank he had felt the pulse of his engines under him, a vibration he had accepted as life, to which his heart beat, his breath-

301

ing was timed. The song of the paddle wheels, that was what was lacking. There was his restlessness, his inability to read or to chat. He, a sailor man of tall masts and good hemp cordage, had been caught by the magic of the engine. What would the off-shore fleet have to say to that? The next thing he knew he would belong to this endless river with its perplexities strewn along every mile and the splendid newness still fresh upon it. Well, what if he did? Worse things could happen. Porter smiled as grimly as a commanding officer should and found his spirit soothed within him. He had tried the impossible and enjoyed it. It was a great war.

Vicksburg was still in the hands of the Confederates, and the only thing the effort had taught the Union forces was that there was no practical way behind Vicksburg. That conclusion was reached by both Sherman and Porter so they gave up the waters of the Yazoo and followed it out to the Mississippi, to Grant's headquarters.

General Grant was fretting at delay. These two, Sherman and Porter, were to have gotten through with boats. Instead they had simply frittered away time uselessly. There was nothing to be gained by further shelling of Vicksburg, except as a matter of harassing the bottled enemy. It was time for attack. The offense was to come from the south. The gunboats and transports alone could get the army from the west to the east bank and instead of being below Vicksburg where they could serve they were blocked by the Southern guns, the threat of annihilation the battery crowned bluff held.

Porter had failed in a gallant effort to do the impossible. He reported that fact to the general. Grant's eyes studied him. The man was talking to no point. He did not understand. Grant wanted the boats below the city. They were still above. He was short of words, but pointed in his remarks. As a commander he was little concerned with the trials of the ditch-skating gunboats. His career was at stake. Vicksburg had held him off too long. It must fall.

When Porter came out of that council he was the soul of the navy, and the navy was sailing very free. He looked down at the river, the Great River. It was the thing that had misled him, that had carried him back of beyond, only to disappoint. Yet it had helped him twice. Had not the waters of the Yazoo, its tributary, risen to clear his stranded boats from the tangled willow shoots? Had not the current aided his hulls in retreat, given them the power to butt their way out toward Sherman through strewn logs, and the speed to hamper the accurate movement of Confederate riflemen? It was not like the sea, but it led to the blue waters of the gulf. Ships there were honest frigates, not mere hulks with engines in them, beating their rythm into a man's soul, but tall sticks and long spars spreading snowy canvas. Far down the river they waited, those free, bird-like ships of the heart. There was no breath of fresh mud there, nor the stain of water-borne silt.

Deep sea ways waited, navy ways, a trim life of exactness and precious routine, long days afloat. Below, toward the heights of Vicksburg, led the river, and beyond. There were no armies on deep water. He had enough of armies. Would the river help him once again? Would it?

There was nothing scatterbrained about Porter. Urged on by failure, by Grant's impatience, by the willingness of the river to bear him onward, he organized his effort well and without undue haste. The transports were bulwarked by bales of cotton, the ironclads fully prepared to the last shell. Then, when every emergency had been thought of, and all had been guarded against as well as they might, with fire hose rigged, dressing stations in waiting and guns shotted, the Union fleet slipped out into the open water amid the summer fullness of a June night. Porter was on his way to the sea.

At midnight, Admiral Porter had established contact with the first of the batteries. Grey shadows, his fleet stretched behind him, steaming quietly without lights, with gun ports closed and every man at his post. Suddenly a single battery opened on Porter's *Benton* which led the column. Under full steam the gunboats sprang ahead. The ports went up and every boat poured out as deadly a fire as it could. There was no attempt at control. Everything outside of that slender, ghostlike column, was the enemy. Shoot where they would, so long as they sent their missiles safely to the

304

THE PORTS WENT UP AND EVERY BOAT POURED OUT A DEADLY FIRE

shore, there were some of the enemy waiting to receive them.

Around the bend at the head of the long peninsula the fleet held on. River eddies toyed with them. Pilots spun their wheels rapidly trying to keep control in spite of the chaos of gun fire, and the shocks of the cannon upon the hulls that held them. Smoke wrapped everything, adding to the bewilderment of the seamen. A vessel hit, spun from its course and left formation. Before he was able to get his bearings the pilot of the unfortunate craft might find he was steering his boat up instead of down-stream. The principal business of the squadron was to get below the forts, but they fought manfully by the way.

Once the column was clearly sighted by the aid of huge bonfires that were lighted on the beach, below the bluff, the guns of the great citadel let loose in pandemonium. The heights belched red bursts of flame. Mortar shells fell in sweeping arcs among the boats. Solid shot from the big rifled guns whined down the stretch of water. Musketry hummed and rattled, good black powder cartridges roaring away heartily, flinging their slug-like bullets at rocking shadows in a river of fantastic water spouts, whirling eddies, lighted by red and orange flares. The cheers of the boat crews were drowned by the thunder of the great batteries. All the boats were hit, some often. The *Benton* was shaken to her very timbers. Bedlam weighed down upon the ear drums, deadening the senses. For months the Vicksburg batteries had prepared for that night's work. For two hours they poured down upon the fleet every projectile they could load. Plates were split, the decks were a shambles, hulls were sprung and twisted, but the Great River had cast its die. Porter had twice been its favorite. Even a vessel that was injured was swept on by the current until a line could be gotten to her and she turned into a well-handled tow. By five o'clock it was over. The heights were quiet again and the dawn came as usual to beleagured Vicksburg, but the Union fleet had passed. Not a vessel had been left behind.

When daylight came Porter sent for his officers to come aboard. Every man had a tale to tell. Every vessel had been battered and wounded, but when Porter mentioned that any boat not fit to go into action would be left behind while the fleet reduced the Confederate works at Grand Gulf every officer began eagerly to belittle the damage that had been done. Nothing serious was the matter. They had been shaken, true, but they were better. Every vessel was more able to fight than before. And so they were for the facing of Grand Gulf which was no mean feat, yet the navy did it in short order in a fast and furious cannonade in which the *Benton* was hit more than twenty times, the *Lafayette* more than sixty and the *Tuscumbia* almost as many. The works were kept so busy that they had no time to devote to the transports which, with Grant's troops on board, crept slowly past and began the enveloping movement of Vicksburg which ended in the surrender of the city on July 4, 1863.

As for Porter, he had retrieved the fiasco of his ditch travel. Those who felt the navy was skulking at the expense of the army had not a leg to stand on. Porter successfully took his iron-clads past the most dangerous fortifications the world had ever seen. What was more, he had taken them through safely. The Mississippi had kept its word with him. He gained fame. Grant gained his army and the Union forces won Vicksburg. All that had come from trusting the open river, taking its force as a gift freely proffered, and not to be churlishly withdrawn.

And the Mississippi? Did it care for the hazards of life and death that night's fighting had offered to the struggling warriors? Did it care for the valor of Porter, or was he a water being of kindred blood and to be humored, to be cajoled? Was it protection or accident that rewarded the clumsy iron-clads? The river never told, but at the last, when Porter had seen the fall of Vicksburg and had cleared up the Confederates on the Red River, he passed by courtesy of the Mississippi down to the Gulf, and so to his frigates and the smell of oakum and tar, finding salt tang in place of the perfume of the river banks in June when the Mississippi played him for its favorite.

CHAPTER FOURTEEN

LEVEES, JETTIES AND MEN

FOR years and years the Mississippi has kept its secret, a brooding thing of zealous strength. Five thousand years ago it began to build a delta for itself that its waters might debouch into the blue gulf. Its reasons were its own, its ends undiscoverable. Proudly racing down, it worked mysteriously at its purpose. Since the days of La Salle its bed has changed so much that most of the route he travelled is now dry land, well back from the river. Still the stream works proudly urged by unknown ambition. In a moment of wrath it overwhelms twenty thousand square miles of land, swallowing the life of it, breaking the hearts of men, leaving ruin and havoc behind. A ravening monster could not be as deadly nor so careless of its strength. There is no understanding the Mississippi.

From the rocky upper stretches it swings down. Sand bars are cut away and rebuilt at will. Savage tree roots, great snags are dropped in a hole in the river bed, dallied with for a day or a month, and then swept on to new lodging places. Banks are undermined, cave in, and the new banks are besieged at once. Islands go down in a night. Some narrow peninsula is attacked by a rise of water, and in a few hours there is a turn gone out of the river, a new island formed and the full body

308

of the stream is pouring through the crevasse, scouring a fresh channel for itself. If the difference in level between the top and bottom of the cut is enough to give the current the velocity of six feet in a second, stratified rock will be broken and washed away by the remorseless strength that never grows weary. At forty inches a second the current sweeps along in its path pieces of flint large as an egg, and even at eight inches a second, a rile of gravel as large as peas, fine sand and clay are carried toward the mouth. If the madness is then upon the river, in its flood state, it empties two thousand cubic feet of these scourings into the sea every second of the day and it heeds neither light nor darkness in its rage. So furiously has its passion been at work that through its lower stretches it runs in a flume of its own building. Its bottom, its banks, the surface of its waters are sometimes all three higher than the country through which it is passing toward its delta. There is no knowledge that can divulge the waywardness nor the wonder of the Mississippi.

Men came to its valley to till the soil, but the river would give them no encouragement. Instead, it drowned them when it could with its floods. These floods knew not time, nor any season. If the rainfall were heavy the river rose, washed over its banks and the damage was done. The Mississippi had extra power then, its life was at its fullest. Some years there have been three, some years not any, but every flood has done damage to some one. The people of the river were not craven. Instead of fleeing from the damage they took it as a matter of course, just another surprise to be added to the score of those already encountered, but when their fields were cleared and put to work, when the centre of both the wheat and corn growing of the country came to be in the valley, then mere brav-ery was not enough. They faced the river in struggle, puny creatures to rise against its power, but determined. The day for building the levees had come.

There was much experimenting. The purpose of the levees was to keep the river from overflowing its shores at high water. They were really auxiliary banks to hold the river when the floods threatened to inundate the country. At first, almost anything was built into the wall of earth. Old stumps, tree trunks, brush, anything to fill it up. The levee was lifted to different heights in different sections. At New Orleans floods raised the river twelve feet. About a thousand miles up, at Cairo, flood water reached fifty feet above normal. At Cairo the river is a mile wide, at New Orleans scarcely half a mile wide. Between the two cities there were places where the river reached out to a forty mile width, if the flood were a bad one. This the levees hoped to stop.

Engineers studied the matter. In general they found a levee with trees or roots in its construction, once the flood topped it, was soon penetrated by the water which worked down along the buried wood and broke through the bottom of the wall. They found the best form of levee to be of narrow top with long slopes leading up to it. Roads running along the summit kept the surface cut up and displaced the clay with which it was faced.

Cattle grazing on them did the same harm. To keep out cattle, some engineers planted orange osage, a hedge shrub, on both the side toward the river and that toward the shore at the foot of the slope. The embankment was also sodded with Bermuda grass. The most general, but least preferable, material was sand. Loam came next in desirability, but best of all was blue clay. This earth weighed more than twice its bulk in water, and besides was impermeable. There was nothing new in the idea of levees. The people of Babylon had used the embankment and the Romans. London, the city of the lake, had once to solve this problem of keeping the waters of a river where they belonged. None of these had to contest with the Great River. There was only one Mississippi in the world and that was untamed. The cost of the levees was well nigh prohibitive. People lost their land by forced sale in order to pay the levee tax they could not meet. Whole counties were poverty stricken, but the levees were undertaken and added to as opportunities permitted.

The floods did not regard the levees with a great deal of respect. Either they began to occur oftener or a better record of their encroachments was kept. From 1857 on the high waters came fast and thick. In 1882, after many tries, the river tore two hundred eighty four crevasses through the walls, desolating the country. The men beside the river repaired the levees and 1883 saw only two hundred twenty-four, while the next year the flood broke through in only two hundred four places. Had the river moved at once to the attack, it might have ended the hope of men that the levee could be made to serve, but it faltered. Ah, truly, none knows the Mississippi! It came not again with its old strength, nor did the waters reach their early might until suddenly they broke loose in 1897 and put a full twenty thousand square miles under water. Another lull followed, and in 1912 the worst outbreak on record came. $42,000,000 wiped out! The whole midland a place of misery and sorrow! The Mississippi was in fine barbaric form, performing at its best, unbridled and proud. Only last year an appeal for help went up from the river people. The Great River had flung abroad its rage in league with wind and rain. What a heart the river must have, what caverns of dark purpose, what wells of strength, what high belief in its mission? Or else it is a dead thing, is that it, the great waters a blind force of nature? No, not the Mississippi, it has a spirit, never fear, as surely as there were once gods on Olympus, or dryads and nymphs in the woods of Hellas. Its mind, that is as inscrutable as death itself.

It was not until the fleet of a thousand steamboats had yielded to the railroads that the river came into its own at the hands of the nation. Then money was spent to make it a worthy water way. Alas, the great boats never profited by it, they had made their last flying trip, flaunted their last brave bit of gilt and crystal, passed in the way of all things from their moment into quiet. When the Civil War was over government funds were raised to undertake river work. Money as a weapon attacked the river, a new enemy, warfare in earnest.

THE RIVER TORE THROUGH THE WALLS DESOLATING THE COUNTRY

On the upper river there were rocks to be gotten rid of, sand bars to be made passable, snags to be dispossessed. Contractors agreed to remove them at so much a snag. Steamboats went to work. In order to make money they cleaned up the river bed, but the obstacles did not wash down fast enough to make the effort profitable so they took to dragging stumps out of the banks. Still their profits were not enough, so they quit. The first rise loosened other snags which washed down and the river was scarcely better for their efforts. River pilots spent their time dodging the new crop. To be effective such work must be continuous, so the government put on its own boats. Engineers were furnished by the War Department. Constant patrol, during the months the river was without ice, was maintained and the work became worthwhile, but the myriad of steamboats it could have served were bleaching their bones on sand bars, or where buried deep in the mud or finishing their days as barges at the end of a tow line.

Sand bars were handled by the Dodge dredge. This was twenty feet wide and was fastened to the bow of a steamboat by heavy chains. The steamboat ran up river to a sand bar, lowered the dredge to the bottom, and as the current carried the boat down-stream, the dredge snatched and dug at the bar, the loosened sand being carried off by the force of the current. When the steamboat had reached the foot of the reef, the dredge was raised and the boat ran up to the head of the bar, lowered its dredge upon a new swathe and repeated its operation. The sand thus removed settled below the bar where it was scooped out and carried away. This method proved effective and gave much credit to its inventor, Colonel Dodge of the Corps of Engineers.

Removing rock obstacles was more costly, but it was done by an expenditure of money. The river cut the open channel ever deeper, growing in strength and value as a highway, but unfortunately the need for the river had departed. The steamboat had passed its glorious zenith. The Mississippi alone profited by the work, save for a few lumbermen who rafted their logs to market down the improved giant. Army engineers thought steamboat pilots ignorant of all but the ability to turn their wheels. Pilots, especially those at some of the rapids, found the engineers taking their livelihood away from them. They resisted. Petitions were sent to Congress asking that the work be stopped, but to no avail. Dikes and wing dams were built by the hundred. Hints, they were, to the Mississippi how to behave. Sometimes in a single night of old savagery the huge stream ran amuck, washing away ten thousand dollars' worth of work, just to prove its strength. Steamboatmen nodded their heads then, rather pleased at the riotous freedom of the old waters. These engineers did not know everything after all. If they had been asked rivermen could have given warning there. Finally, the engineers took some of the practical experts into their councils, sharing the work to a degree with those of local knowledge. Results were better then, the river was curbed and controlled more surely, money was saved and opposition to the work died away.

ﾠThe course of the river grew better as the years went on. From Cairo to the mouth it runs between long lines of levees. Much of the work was bad and had to be replaced. When local means had failed the national government took over the work. Concrete and sheet piling have been added. It cost money to build those first brave levees. It cost much more to keep them in condition where they could do battle with the river, but in general the levees have been successful. River people live closer and more safely beside their strong master than ever before. As the river bed rises the levees too must rise. It is an endless war but a good one, a man's war with the uncertain temper of the most important river in the world. Even the Mississippi must respect that which it glowers upon, that which in its heart it hopes to tear and ruin in a wild moment of freedom, when the wind and rain help it to do so.

At the river mouth the true Mississippi stands revealed. It was there, where the whole eager burden of its stream goes pouring into the sea that man made his bravest stand to thwart its desires. There are three principal passages into the Gulf. On the east it sweeps out through the Pass à l'Outre, on the west through a curving channel known as the Southwest Pass, and in the middle the narrow but straight South Pass. So great a river should bear the world's freight to the sea, but by years of labor it had instead carried down a great burden of sand and clay, and wearied of its zest, dropped it where the current slackened and the Gulf of Mexico swallowed the giant's being. No easy matter this for mere men to overcome, the desire of the river to be rid of its burden. As early as 1837 a group of army officers of that quaint and old fashioned sort that were so common in our early military service weighed the matter. First they recommended that the bottom of the mouth be kept stirred up by dredges, chains and scoops. Perhaps the river could be prevented from becoming so weary and would carry its silt on out into the Gulf. If this failed it could be bucketed out. A fine dream that, but one that despised the Mississippi. They had underestimated the power of the river, betrayed perhaps by the fact that at New Orleans the river was only a half mile wide and one hundred sixty-eight feet deep. In event of being unable to win by buckets, the silt was to be hurried on its way by building of long jetties, narrowing the river mouth and thereby increasing the current so that the river itself would be made to carry the load of sand and clay into the Gulf. If all failed they were willing to build a ship canal and give up going to sea by the river itself. So they spoke, but little was done. The river was waiting for a man as full of faith as Pére Marquette, as bold as De Soto, as relentless as La Salle and as brave as Jackson. It waited for years, years of futile work, mere puddling and experimentation.

There was a boy who in 1837 was caught by the spell of the river. He read widely. His life was a tale come true. A lad of fourteen, in straitened circumstances, he spent his days as a clerk, but at night he had access to his employer's library. They were busy nights, spent poring over engineering books, spent at study of all that mechanical philosophy had to offer him. When the strain grew too great he could step out under the stars and listen to the river suck and growl along its banks. It was a challenge to him that lightened the long hours at his desk. Thought upon in the dark, the river called for battle, but first there was much to learn, mysteries to explore, truths to master. The early life of James B. Eads was a busy one, without idleness, and disciplined by hard study. Alone he was able to acquire what many would have failed to get at the hands of a good teacher. He was not alone in spirit. It was a day when sixteen year old boys were taking ships around the world, and were commanding gunboats up many rivers. Fourteen was a golden age to begin his battle.

A day came when he left the study to come to grips with the river itself. Many a boat was lost in river trade. A shifting bar, a halting snag, a falling river, and a steamboat passed from safety to a hulk on the bottom. Eads went into wrecking, the business of examining steamboats as they lay on the river bed with an eye to raising them and putting them back at work, plying their trade. The diving bell was the most common device used in those surveys, whereby a man could get below the river water and see for himself the plight of the boat in question. It was the chance of a lifetime for Eads, for he saw not only the boat he was raising, but the river itself.

Under the diving bell he trod the river bottom. At high and low water he studied the current, the formation of the bed at hundreds of places along the river course. The sediment the stream carried, the speed with which the huge power worked, its manners and customs, these he saw at first hand. Reflecting them against the theory his study had given him he made much shrewd use of them, accurate observation that led on to hopes and fears, the background of understanding. He was making the river his.

The Civil War saw him at work building seven gunboats, primitive iron-clads for his river. He turned them out in a hundred days. When he began, the trees he needed for lumber were still standing, and the iron still unwrought, but he got his boats out in his hundred days. Then he saw the uselessness of using timber for backing and built six others wholly of iron, a brave departure. In addition, with the increase in the size of guns, the difficulty of handling them quickly and accurately had increased to a point that gave the gunners an almost impossible task, which Eads cleverly relieved by applying steam to the problem. It was thus he laid the foundation for his work as an engineeer.

The war ended at last. Many men emerge into the limelight at war time, only to drop back into obscurity in peace. Their ability was one sided. That of James B. Eads was many sided. He stood by his river. He knew that at St. Louis there was much quicksand. He knew that below it somewhere was bed rock. When the need came for a bridge to span the Mississippi at St. Louis he undertook the work with the same quiet assurance he had used in probing the river bottom in his diving bell and building his iron-clads. He went down through one hundred feet of quicksand and put his piers firmly on bed rock. Since he could get no foothold for the crib work of his staging, to hold the steel arches till they were completed and able to hold themselves, he found a new way of his own to lead the steel skeleton out across the water until his bridge was at last permanently flung across the river. From pier to pier, the span was five hundred feet. When success came the position of Eads as a great engineer was assured. The nights along the star lit Mississippi had borne fruit. He was ready for a real battle, the war with the river's full strength at its mouth, and the nation was ready too.

For twenty-five years the government had looked and talked, but had done next to nothing. In 1873 a Congressional delegation took Eads down to the mouth to look the matter over. He declared himself for the building of jetties. It was folly to try to bucket out the sediment, a ship canal was not necessary, stirring the bottom was inadequate. The Mississippi could not be stemmed. It had to be put to work, but the force, the power was there. The time had come when Eads could grapple with a work for giants. He was surrounded by jealousies. The United States Engineer Corps felt such work was their province. The City of New Orleans wanted a ship canal, because they felt that the city would develop more fully and be more than ever a transfer point, through which all the valley travel would pass to the great outer world. To Eads, there was the river he knew and loved. It was his life, this stream and he knew its bent and desires. If ever human being understood the foibles of its heart, James B. Eads was the man, chosen in boyhood to be the high priest of the fabulous monster, prepared for the effort by years of patient thinking and brutal, man killing work. Man the tiny animal against this mighty natural force! He thrilled to the opportunity. When the Congressional delegates listened to him he noted their respect, their deference. He had them in his hands. The river waywardness, rival influences, envy, the obstructions that government employees could thrust in his path, all these he accepted gladly. He would give the Mississippi an open gateway to the sea. It was so magnificent a thought that he could hope to find its equal no where else in the world. As an engineer he knew that jetties would do the work, hurling the whole force of the river against the bars at the mouth, forcing it to fight for its union with the sea instead of allowing it to spread its energies in a shoal, fan shaped flow, a waste of power unworthy of its destiny. In 1770 only three feet draft could be carried to Glasgow. By jetties and wharves, in a hundred years, the Clyde had deepened itself so that ships drawing twenty feet docked there in 1870. The Clyde was a rivulet compared with the Mississippi. Eads began to work upon the new idea. He had learned to organize on the St. Louis bridge, but even that wonder was puny compared with this. He took a year to get his facts in orderly array. He considered all phases. Then he opened the fight for his river.

In 1874 he formally presented to Congress a plan

for opening the mouth of the river to a depth of twenty-eight feet. All risks he and his associates assumed. The work was to cost ten million dollars but not a dollar was to be paid until a depth of twenty-three feet was secured and then one million dollars for each additional foot. When the depth of twenty-eight feet had been obtained the government would still owe him five million dollars according to this plan. As a pledge that his work was permanent the remaining money was to be paid over ten years, at five hundred thousand annually. It took faith to make a pledge like that. Eads was taking every risk. The arrangement was more than generous, but the influence to be combatted was great. One might have thought Eads had opened a war instead of having offered a practical plan for one of the greatest engineering works of the age. It was argued that his plan was no plan, that it was not practical. Letters were written to the bulk of volumes, pro and con. Never did the wheels of progress find a rougher road, but Eads was equal to it all. He had taken a year to prepare and it was enough, yet first there would have to come delay. A commission was sent to Europe to study what had been done there.

They gazed upon the dykes of Holland, designed to withstand action of surf as well as the thrust of currents. They brooded upon the works of the Dvina, a Russian River, bearing sediment into the Gulf of Riga. They studied the Rhone, where jetties had increased the depth over the bar from five to thirteen feet and the Danube where depths of nine feet had been increased to twenty. In every case they were anxious to learn if the river borne silt was really carried out into the sea or dropped at the river's end. Would it only increase the size of the bar, killing all the gain in channel depths? Eads had established in his own mind, while working with the diving bell that sand and silt were carried in suspension and not pushed along the bottom. He believed, that given current enough, the river would carry its deposits far off shore and clear the river mouth. The jetties would narrow the stream and increase the velocity of the current. For him there were no doubts. When the commission returned there were no doubts for it either. Congress passed a bill authorizing the work, and construction began in 1875.

Eads financed the matter skillfully and then gave himself up to the struggle with the river. He considered the three finger-like passages reaching toward the sea and chose the middle for his effort, the straight, if narrow, South Pass. The Government engineers thought his choice bad. For years they had toyed with the Southwest Pass as the best passage to keep open. Eads was not to suffer for lack of critics. At first things went very slowly. In June 1875 there was but little hint of the bridling of the river. A small white flag here, a temporary scaffold there, a drift log imbedded in the mud and topped by a stake, a spindle shanked tripod astride a mud lump, an instrument stand upon a reef: none of these things gave promise of the magic about to be realized. Then down came the contractors. Steamers, tugs, dredges puffed about. Pile driving gangs went to work. Willow cutters waded into swamps. Blacksmiths set up their forges and the tinkle of beaten steel went out over the water. From the start sickness entered the gangs, and a resident physican was both present and busy. The jetties were under way.

Just what are these jetties? How was the fate of the river to be altered by them? In Mr. Eads' own words "The improvement of the mouth of the Mississippi proposed by me consists in an artificial extension of the natural banks of one of the passes from the point where they commence to widen and disappear in the Gulf, to the crest of the bar, about five miles distant." An engineer's own statement that, short and to the point, edged like a good tool, worthy speech. Yet it was easier to state the matter than to achieve it. The jetties were built on several layers of willow mattresses sunk by stones piled upon them. The willow kept the jetty from sinking into the sand, and the sand, as it partially swallowed the willow protected the wood from attack by the marine worm that ruins unprotected boats in a month in southern waters, the teredo. The upper works were topped off by concrete blocks. Ingenuity, back breaking work, temporary failure, hindrances of a thousand sorts had their part. Eads and his contractors mastered them. There was no turning back. A government engineer, disgruntled over the choice of the South Pass, undertook an official survey, under his commanding officer, to prove Eads was not succeeding. He notified both New Orleans and St. Louis newspapers there was a depth of only twelve feet in South Pass. Mr. Eads refuted this by a statement that there was then sixteen feet of water on the shoalest bar, and that was deepening rapidly. Like wild fire the two reports went over the country making no end of a row. Rivermen took sides. In addition to finding his engineering work at its most critical and exacting stage, Mr. Eads was involved in financial troubles and embroiled with the government through the misrepresentation of the envious military engineers. The jetties needed a friend who had faith. They were already working night and day although incomplete. The river was responding. Wanted—a friend with faith and courage.

Captain E. V. Gager of the steamer *Hudson* had long believed in the South Pass. He believed in Mr. Eads. He believed in the Mississippi's power to use the jetties and he desired to put his faith to a test. On May 12, 1876, the centennial year of our nation, Captain Gager bore up for the South Pass. The *Hudson* was two hundred eighty feet in length and close to two thousand tons in burden. She drew fourteen feet, seven inches. If what Eads' enemies had said were true the ship would pile up in three feet of mud, for the tide was not full, but falling. If Eads had spoken truly he would be able to telegraph the *Hudson's* passage over the country. It was public vindication, or for Captain Gager perhaps a lost ship.

"Head her for the jetties," was the order.

"The jetties, sir," said Pilot Francis. The captain rang for full speed, and with a bone in her teeth the proud bow pushed past the head of the jetties. She held her way, never scraping the sand. On and on, until in thirty-five feet of water off Port Eads she checked her speed. She was the first ship of anything like her size to risk a passage. Captain Gager had proven his faith. Mr. Eads had a friend. Confidence in the jetties was restored and loans were no longer hard to make.

Work on the jetties leaped ahead after that. Red dirt from the plateau of Cape Town, gravel from the English Thames, granite from Rio de Janeiro, sand from France and Spain were poured into the building. Every ship coming in ballast unloaded at Port Eads and went up to New Orleans light, ready for its cargo. Shipping using the passage annually approached a thousand vessels with a maximum draft of twenty-four feet, six inches. In 1878 the work was ended with a depth of thirty feet. From then on the task was to maintain and build auxiliary works to keep what had been won. The government engineers fought to the end, still trying to disparage the results when final payments were being made, but Eads had established the use of the jetties to keep the river mouth open. He admitted it might some day be essential to jetty the Southwest Pass also and so it came to be. That was twelve years work under modern conditions, finished in 1912, but Eads remains the champion of the river, the man who dared, who fought and won.

And the river, what does it think of it all? As a man

stands and looks down from the top of the South Pass
Lighthouse above the flat marsh of mud, reeds and
grasses can he read the heart of the tide that owns this
land? At the far east loiters the low, willow-fringed
banks of Pass à l'Outre, a dim pencil line between the
blue waters of Garden Island Bay and the sky dome.
To the west is the arc of the Southwest Pass, flat against
the horizon. The South Pass Light towers a hundred
feet in the air, a white, pyramidal skeleton tower. A
few fishermen live close by, but it is the last outpost
toward the Gulf, a sentinel upon the jetties.

To its foot swings down the South Pass from ten
miles away, ten miles of pleasant channel, a thread of
clear water through the yellow swamp. On the higher
grounds are dense reed growths, sea cane, and on the
lower, coarse grass, and everywhere the sand is spotted
by small bays and passes that lead nowhere, water and
mud that is quick and bears no weight. There is not a
shoal, nor a sharp turn, nor a dangerous bank in the
whole of the South Pass and down it speeds the Mississippi as heedlessly as it flows through its other passes to
right and left flooding its intricate bayous, its crevassed
swamp banks.

Who knows its heart? Did La Salle, or Marquette, or
Cabez de Vaca, or Mike Fink, or Murrell? Is it rogue
or honest, did it speak to the beating paddle wheels or
echo its true self to murderous pistol shots? The arms of
it reach down under the sky from the mountains and
seize upon the blue Gulf under a coppery sun where

there are none to know of its heart, where there is silence save for the voice of the great waters that mix and mingle, quiet at the last, done with its thousand cities and the tales of men.

THE END